BEHIND LOCKED DOORS

Pietro Perugino (1448–1523). *Delivering the keys of the kingdom to St. Peter*, ca. 1481–1483. Sistine Chapel, Vatican Palace, Vatican State. Copyright © Scala/Art Resource, New York.

BEHIND LOCKED DOORS

A History of the Papal Elections

FREDERIC J. BAUMGARTNER

First published 2003 by
PALGRAVE MACMILLAN™
175 Fifth Avenue, New York, N.Y. 10010 and
Houndmills, Basingstoke, Hampshire, England RG21 6XS.
Companies and representatives throughout the world.

PALGRAVE MACMILLAN is the global academic imprint of the Palgrave
Macmillan division of St. Martin's Press, LLC and of Palgrave Macmillan
Ltd. Macmillan® is a registered trademark in the United States, United
Kingdom and other countries. Palgrave is a registered trademark in the
European Union and other countries.

ISBN 0–312–29463–8 hardback

Library of Congress Cataloging-in-Publication Data
Baumgartner, Frederic J.
Behind locked doors : a history of the Papal elections / Frederic J.
Baumgartner.
   p.   cm.
   Includes bibliographical references (p.) and index.
   ISBN 0–312–29463–8
   1. Popes—Election—History.   I. Title.

BX1805.B324   2003
262'.19—dc21

                    2003043306

A catalogue record for this book is available from the British Library.

Design by Letra Libre, Inc.

First edition: September 2003
10 9 8 7 6 5 4 3 2 1

Printed in the United States of America.

Contents

Acknowledgements

I must acknowledge my debt to the Interlibrary Loan staff at Newman Library at Virginia Polytechnic University and State University; without their extensive help this book could not have been written. My colleague David Burr provided an expert critique of the work and checked my translations from Italian. Andy Nichols, my research assistant, was most helpful as a fact checker and critic of my style, although any errors that remain in the text are strictly my responsibility. And I must express my appreciation to Michael Flamini and his fine staff at Palgrave Macmillan/St. Martin's Press for their efforts in seeing this work to publication. And as always my family, especially my wife Lois, has supported my efforts at authorship with good humor and patience.

Preface

By the time the reader picks up this book, it is quite likely that the exquisite drama of a papal election will have already taken place, considering the age and ill health of John Paul II, the current pope. For that reason I do not make any effort to predict the outcome of the next conclave, other than to make a comment about St. Malachi's prophecies, which supposedly have identified all the popes since 1143. My purpose is exploring the history of the papal election and showing how it evolved into its contemporary mode. I wish also to demonstrate how modifications in the format of the conclave have resulted in changes in type of men elected pope, who have heavily influenced the nature of the governance of the Catholic Church through the centuries and more broadly the history of Western civilization. As the oldest system by far for choosing the leader of an institution of any sort, the history of the conclave demonstrates how legitimacy can be achieved and maintained despite the personal foibles of those who have been chosen pope and those doing the choosing.

Since this book is intended for a broad readership, I had planned to base it on secondary sources such as Ludwig von Pastor's forty-volume *History of the Popes from the Close of the Middle Ages,* which has the only account in English for several conclaves from 1417 to 1774. G. Zizola's *Il Conclave Storia e Segreti: L'elezione papale da San Pietro a Giovanni Pietro II* (1993) is a valuable history of papal elections, but well over half of it is devoted to the last five conclaves. In English the best work devoted strictly to the conclave is Francis Bickle-Young, *Passing the Keys: Modern Cardinals, Conclaves, and the Election of the Next Pope* (1999). As the title suggests, it deals largely with the twentieth century, and it has little on earlier conclave history. Most papal biographies and histories of the papacy contain information on the papal elections, but too many authors of such works are either pious hagiographers or scandal-mongers, far too willing to accept rumors for good or bad about the popes and the conclaves that elected them. A good example of the latter is Valérie Pirie, whose *The Triple Crown: An Account of the Papal Conclaves from the Fifteenth Century to the Present Day* (1935) is a popular account of the papal elections from 1455 to 1903. The author eagerly reports every rumor and salacious story, making the book highly amusing but hardly accurate in its account of conclave history.

The many discrepancies and gaps appearing in the secondary sources led me to seek out primary sources in the Archivo Segreto Vaticano. The vicissitudes of working there, with its arcane rules and restricted access, made my time at the Vatican less productive than I had expected. I also found that less material from the conclaves had been preserved than I hoped. Yet the trip provided me with important information for the conclaves mostly from 1600 to 1846. The greater emphasis on secrecy, beginning with the 1878 election and increasing steadily to the present, has reduced the accuracy of information about the recent conclaves. One irony of studying the history of papal elections is that there is far more first-hand source material for the conclave of 1549–50, for example, than for the most recent ones.

Although the issue of papal finances is only a tangential concern in this book, a brief description of papal money is in order. In the late Middle Ages the popes used the ducat, a gold coin modeled on Venice's, as their principal currency. In the sixteenth century gold crowns, *scudi,* from the French coin the *écu,* became the standard currency. In the seventeenth century five *scudi* were worth one English pound sterling, and in 1820 one was worth 5 francs 37 centimes in the French money of that time. During Pope XI's reign it was dropped in favor of the Italian *lire,* which was replaced by the euro when it was adopted in Italy in 2001.

I have included an appendix devoted to explaining the origins of what is certainly the best-known aspect of a papal election, the use of smoke from burning the ballots to signal the result of a ballot. I have kept the notes to a minimum, using them mostly to cite direct quotes or to indicate when there is a disagreement in the sources over a significant point. Some notes, however, do include annotations on the text. The "Other Sources Consulted" section does not include those works already cited in the notes. A glossary of terms used in the book is placed at the front of the text so that the reader might peruse it before beginning to read the book.

Glossary

Accession, a former procedure of allowing cardinals to switch their votes after the formal written ballot to one of the stronger candidates in hopes of immediately ending the conclave. Until 1621 it was done verbally; thereafter, there was a written ballot of accession.

Acclamation, a former procedure for electing the pope that involved the verbal assent of most if not all the cardinals to a candidate. It occurred only after there had been one or more formal scrutiny. In the Middle Ages it was called "Election by Inspiration."

Adoration, a former procedure for electing the pope that involved the verbal assent of all the cardinals to a candidate, which took place before any formal ballot.

Antipope, someone chosen pope by a faction of the cardinals or by a Catholic prince, and who does not appear on the official list of popes published by the Vatican.

Bull, a formal papal document, so called from the *bulla* used to seal it and make it official.

Camerlengo, the papal chamberlain, who has the responsibility of seeing to the funeral of the dead pope and arranging the conclave.

Capitulation, formerly an agreement written at the beginning of a conclave setting forth the relationship between the next pope and the cardinals. All the cardinals would swear to observe it if elected.

Cardinale-padrone, "cardinal-master," term used in the ancien régime for the pope's nephew, named a cardinal quickly after his uncle's election, who usually had the office of secretary of state.

Conciliarism, the belief that the general council of the Church was superior to the pope and ought to have a role in electing him.

Conclavist, an aide to a cardinal in conclave.

Consistory, meeting of the pope with the cardinals. In the Middle Ages it occurred frequently and involved a wide range of church business. Since 1600 the principal business is announcing new cardinals and bishops.

Curia, curialist, terms for the papal court and those with offices in it.

Exclusion, a former practice of the major Catholic rulers of vetoing cardinals whom they believed hostile to their interests.

In petto, Italian for "in the breast." The name of a new cardinal is kept secret. If it is not revealed before the pope dies, he never joins the College of Cardinals. The next pope has no obligation to give a red hat to those whose *in petto* nominations have not been published before his predecessor's death, should he learn who they are.

Interdict, a ban on giving the sacraments, imposed on a country or diocese by the pope or bishop for an offense against the Church committed by its government or people.

Nepotism, from the Latin for nephew, it refers to the practice of giving offices to relatives.

Papabile, papabili, "popeable," a term used for the cardinals who are regarded as having a good chance of being elected pope.

Prattiche, an Italian term used to describe the cardinals' discussion about the *papabili* before the conclave begins.

Red Hat, the distinctive broad-rimmed hat that cardinals wore beginning in the 1200s. Its color, the same as for a cardinal's robe, was said to symbolize a cardinal's readiness to be a martyr for the Church. "To be given a red hat" means to be named a cardinal; "raised to the purple," another phrase with the same meaning, is not used in this book.

Schism, from the Greek for tear, it refers to a split in the leadership of an institution, especially a religion.

Scrutator, one of the cardinals who have the duty of counting the ballots during a scrutiny.

Scrutiny, narrowly defined, it refers to the process of determining the validity of the written ballots in the papal elections and counting them; more broadly, it is a term for the process of voting.

Sede vacante, "the vacant see," it refers to the time between the death of the pope and the election of the next one. The College of Cardinals has the authority to make decisions for the Church, but it is limited to only the most urgent matters.

See, from the Latin *sedes,* "seat," it refers to the jurisdiction of a bishop; Holy See is used for that of the bishop of Rome, namely, the pope.

Ultramontane, "across the mountains," used to describe those who had a very strong sense of papal authority over the Catholic Church outside of Italy.

Zelante, Zelanti, from the Italian for zealous, it referred to those cardinals of the eighteenth and nineteenth centuries who wanted to elect popes known for their virtues and their zeal for the well-being of the Church, not their political connections.

CHAPTER I

From Apostle to Pope

No other event in human society can compare with a papal conclave in its combination of tradition, drama, ceremony and pomp, and significance. Once the reigning pope's death is announced to the world, cardinals and reporters rush to Rome from the farthest corners of the globe. As they assemble in the Eternal City, the drama and tension of the upcoming election begins to build. In the evening of the fifteenth day after the pope's death, the papal master of ceremonies shouts: "Extra Omnes!" "Everybody out!" The doors of the Sistine Chapel in the Vatican palace are locked *cum clave*, "with a key," from within and without, and only the cardinals are left to begin the process of electing the next pope. Twice a day over the next several days, the cardinals announce to the world their failure to choose a new pope by burning their ballots with wet straw in a special stove to produce black smoke that rises from an ancient chimney atop the Vatican. It is the focus of attention of an immense crowd in St. Peter's square. Then, as frustration begins to mount, what appears to be white smoke puffs out of the chimney. "Is it white?" The crowd strains to see. "It is white!" "Who is he? Who is the pope?" the assembled thousands ask in one voice. After what seems to be an endless wait, the senior cardinal-deacon appears on a balcony overlooking St. Peter's square and announces: "Habemus papam!" "We have a pope!" Then in sonorous Latin, pausing between words to add suspense, he gives the name of the man elected and the name the new pontiff has chosen. Is the new pope someone whom the crowd recognizes, or is he one who comes as a complete surprise? Soon the new pontiff himself comes to the balcony to give his first blessing *urbi et orbi,* to the city of Rome and the world.

Such is the modern papal election, the longest-standing method of choosing the wielder of authority for any institution. In its setting and ceremony it appears to be thoroughly medieval, but in many of its details it is in fact only a century old, and the next election will reflect major changes made only in the 1990s. Few Catholics or non-Catholics realize that the papal election has gone through 2,000 years of change and development. The present system would be unrecognizable to those Christians of two millennia ago who chose the first bishops of Rome. The Church as a whole has changed enormously since the first century, but few aspects of it have been transformed as much as the method of electing the pope. As the importance of the position of bishop of Rome increased, so did the number of persons who had a stake in influencing the outcome of the election. The modern conclave is largely a result of efforts to eliminate or at least reduce the ability of outsiders, largely the Roman nobility and Catholic rulers, to influence the election of the pope. The history of the conclave is, therefore, truly a political history of the papacy and the Catholic Church, since changes in electoral process must be seen largely in light of the political pressures on the papacy that came from both within and without the Church. A history of papal elections is also a history of the College of Cardinals, since the cardinals have been the electors since 1059 and almost every pope since then has been a cardinal. Thus, to understand the conclaves and their outcomes, it is necessary to note changes in the College's composition and the types of the men appointed to it over the centuries.

The authority of the pope over the Catholic Church draws on the claim that as bishop of Rome, he is the successor to St. Peter, the leader of the apostles and first bishop of Rome. The major biblical texts on which Peter's primacy is based are the listing of the apostles in which Peter's name is first (Matthew 10:2; Luke 6:13); Jesus's renaming of Peter: "You are Peter, and on this rock I will build my church." (Matthew 16:18); and his charge to Peter to feed his lambs and his sheep (John 21:15–17).[1] In the Acts of the Apostles, Peter is the spokesman for the apostles and the Christian community. He tells the community to select Judas's replacement, admonishes Ananias and Sapphira for their deceit, refutes Simon Magus, and makes the decision to include the gentiles in the Church. To be sure, in the second half of Acts, Peter nearly disappears, as Paul, the apostle to the gentiles, takes center stage.

According to church tradition both Peter and Paul died at Rome, but the Bible and secular sources are equally silent on the place, date, and manner of their deaths. From early on tradition proclaimed that Peter arrived at Rome in 42 AD to found the Roman church and was executed there in 67. A later tradition would say that no pope would ever see the number of years, twenty-five, that Peter was bishop of Rome. By 1300 the admonition "You will not see St. Peter's years!" was included in the papal coronation ceremony. It was proven wrong only with Pius IX (1846–1878). Most modern scholars doubt that Peter arrived in Rome as early as 42, but few dispute his death there, although they are more inclined to argue that his death occurred in 64. Rome was the Apostolic See, blessed with the presence and martyrdom of the two great apostles. As Irenaeus of Lyons wrote in about 180, the church at Rome "was the greatest and oldest, because it was founded and built by the glorious apostles Peter and Paul."[2] Rome's role as the capital of the Empire for Christianity's first four centuries also contributed greatly to Rome's primacy within the Church.

Should Peter be counted as the first bishop of Rome and therefore first pope? Early Christians would not have called him bishop. The title of apostle was distinct from bishop, which came from επισκοποσ, "overseer," and referred to those who presided over a local church and saw to correct belief and the needs in the Christian community. The first use of bishop in the Scriptures comes in Acts (20:28), when Paul admonishes the elders of Ephesus to keep watch over the flock "of which the Holy Spirit has made you overseers." In his letter to the Philippians (1:1), Paul greets "the saints in Christ Jesus who are in Philippi, with the bishops and deacons"; and First Timothy and Titus (the attribution of either book to Paul is questionable) list a bishop's qualifications. Otherwise the word is not used in the New Testament. In the writings of the second generation of Christians, the presence of bishops as leaders of Christian communities is clearly established.

There is no notice in the Bible of how bishops were selected, but the process of choosing the seven deacons detailed in Acts (6:1–6) influenced the way they were. The apostles told the community in Jerusalem to identify seven men "full of the Spirit and of wisdom." Those chosen stood before the apostles, who prayed over them and laid hands on them. The earliest non-biblical statement on selecting bishops probably is that found in the *Teaching of the Twelve Apostles* dating from shortly before 100: "Elect for yourselves bishops and deacons, men who are an honor to the Lord."[3] The method of choosing the bishop was quite similar across the Christian communities by 120.

Numerous sources from the second century indicate how the Christian communities elected bishops, but very few refer to Rome. Evidence about the Roman community and its bishops is surprisingly sparse for the era before 250, considering the importance they have in church history. The earliest Roman sources, which date to about 400, propose that Peter appointed Linus as his successor and named Anacletus and Clement to follow in turn. The lack of information on early Christian Rome resulted in the creation during the early Middle Ages of myths about the authority of the early bishops in Rome, which were used to enhance papal power in later centuries. What hints there are suggest that in Rome there were several bishops at the same time or perhaps even no bishop until after 100. The earliest official list of Roman bishops, the *Liber pontificalis,* dating from about 354, is regarded as unreliable for the first two centuries of the Roman church.

Eusebius related a charming legend about Fabian's election in 236. Fabian was not among the candidates whom the assembled brethren were considering, until a white dove landed on his head. "Thereupon the people, all as if impelled by one divine spirit, with one united and eager voice cried out that he was worthy, and immediately they set him on the episcopal seat."[4] Clearly, the choice of the bishop was the public concern for the entire Christian community of Rome. When Fabian fell victim in 250 to Emperor Decius's persecution, the most widespread attack on Christianity up to that time, the Romans put off electing a new bishop for about fourteen months.

From the letters of Cyprian of Carthage, who served as bishop of Carthage from 248 to 258, we gain an insight on the situation in Rome during that time. He knew about events in Rome and had great respect for the bishop of Rome as the successor of Peter, yet he as a bishop regarded himself coequal in authority. During Decius's persecution, factionalism erupted in the Church over the issue of what to do with those who had given in to the imperial command to burn incense to the divine emperor. A rigorist group demanded that they be rebaptized before being readmitted, while a lenient faction only required penance. Both factions elected bishops: Cornelius by the forgivers, Novatian by the rigorists. Both men sought support from Cyprian, who sided with Cornelius. Cyprian wrote to others laying out the reasons why he supported Cornelius as the true bishop of Rome. He stated that Cornelius had advanced through all of the clerical offices of the Church one step at a time (Cyprian himself had become a bishop only two years after his conversion), nor had he campaigned for the office of bishop but had to be persuaded to accept it. Cyprian went on to write: "Moreover, Cornelius was made bishop by the choice of God and of His Christ, by the favorable witness of almost all of the clergy, by the votes of the laity then present, and by the assembly of bishops."[5] He also emphasized that Cornelius had been ordained as bishop of Rome by sixteen bishops of the region around Rome, while his rival had the support of only three. This was the first definite schism involving the church at Rome (there are hints at earlier ones), and it was resolved largely by a display of support for Cornelius from most of the bishops near Rome. Ending the schism in favor of Cornelius helped establish that there could be only one bishop in a city, defined as church law at the Council of Nicaea in 325.

At Nicaea, Emperor Constantine presided over the assembled bishops, while Sylvester I, the bishop of Rome, sent two priests as his representatives. While Constantine, the first Christian emperor, made no changes in the method of electing bishops, his authority added another element of influence into the election process, at least in the major bishoprics. His vast gifts to the Church, concentrated in Rome, also required that the bishop of Rome become far more an administrator of property than had been true earlier. The close cooperation between church and state began to have an impact on the type of men being elected. Constantine, however, had little opportunity to influence elections of the bishop of Rome, since Sylvester I held the office from 314 to 335, the longest for a bishop of Rome at that time and one of only twelve to exceed twenty years to the present. It nearly matched Constantine's own reign from 312 to 337. Sylvester's long term was followed by Mark's short one of only ten months. Mark has a place in papal history for designating the bishop of Ostia, a nearby city, as first among the bishops who consecrated a new bishop of Rome. Later the dean of the College of Cardinals held the title of bishop of Ostia.

Constantine, before dying in 337, had a hand in the election of Julius I, who defended the Nicene Creed against the Arians. The Arian controversy became a serious problem for Julius's successor Liberius after he was elected bishop of Rome in 352. Constantius, Constantine's son, became sole emperor about the same time. As an Arian he was eager to impose that version of Christianity on the entire Empire. Failing to persuade Liberius to accept the exile of Athanasius, the stridently anti-Arian bishop of Alexandria, Constantius banished him and named the archdeacon Felix as bishop of Rome. Felix II had sworn the oath taken by all Roman clergymen to accept only Liberius as bishop, but pressure from the emperor was too strong to resist. After three years in exile Liberius gave in to Constantius and condemned Athanasius, and he was allowed to return to Rome. Most Romans accepted him back as bishop. Felix was forced to flee the city, but he kept a handful of followers. Despite having been exiled for his faith, Liberius was the first bishop of Rome, and the only one until 496, not to be honored as a saint, while Felix was recognized as both pope and saint but now is deemed an antipope.

Both men died within months of the other, but the coincidence was not enough to reunify the Christians of Rome. A majority supported Damasus in 366, while the others elected Ursinus. Both had the support of bishops to ordain them. Damasus had backed Felix until Liberius's return, but he now took a hard line against those he regarded as apostates. Whether or not he encouraged it, his supporters seized the church where Ursinus's people were gathered, killing several. The Ursinians then also resorted to violence, but Damasus's men were more numerous and more bloody. When they stormed another Ursinian stronghold, 137 men were killed.[6] Ursinus fled Rome, and Emperor Valentinian, to avoid further violence, ordered him to stay away.

It is at this point in the history of the Church that the term pope can properly be used for the bishop of Rome. The word *pappas* (παππασ) from which it comes was used for any bishop in the second century, but by 300 it was restricted to the major bishops, and in the West, only to the one at Rome. When the election of the Roman bishop could lead to violence, a common occurrence in the Middle Ages, it is clear evidence that the prestige and authority of the office had reached the level associated with the papacy.

Damasus had won the office of bishop of Rome in a dubious manner, but it did not stop him from asserting claims for his authority that would not have been out of place for his powerful medieval successors. He persuaded the emperor to issue a decree naming the bishop of Rome "bishop of bishops" and giving him final jurisdiction over all church affairs. He antagonized the Eastern bishops to the point that they held a council in 381 to deal with his claims of supremacy. They decreed that the major bishops of the Eastern Empire had the right to administer the Church in their regions. They also accepted a statement that the bishop of Constantinople was preeminent in honor after the bishop of Rome. The stage was set for the schism between the Greek and Latin Churches.

Using the term pope for the bishop of Rome, however, should not be taken as an indication of major change in the method of electing him. The clergy and laity of the city continued to determine the outcome, but politics, both at the local and the imperial level, played a role in the elections. Family ties also were a factor at times. In 401 Innocent I succeeded his father Anastasius I, one of three father-son successions in the history of the papacy.[7] Innocent was pope when the Visigoths sacked Rome in 410, but he was then at the imperial court in Ravenna. At his death in 417 the Roman church chose as his successor a man who did not have roots in Rome, the first bishop since about 100 for whom that was true. Zosimus was a Greek who had come to Rome during Innocent's reign. His election was not contentious, but his reign was. He sparked controversy by deposing several bishops in Gaul, indicating how far-reaching claims of papal authority had become. Zosimus reigned for two years, and the existence of factions for and against him complicated his successor's election. His supporters elected Eulalius, who was also a Greek, while his foes chose Boniface I. While there was no bloodshed, the unrest in Rome forced Emperor Honorius to step in. The emperor at first recognized Eulalius but turned to Boniface, because the latter had more support among the Romans. He held the office until his death in 422, while Eulalius was exiled. The emperor decreed that when there was a double election in the Roman see, the entire Roman community should decide the issue by unanimous selection. No disputed papal election was decided that way, but the decree is indicative of imperial concern over who became pope. Celestine I, Boniface's successor, was elected without opposition, probably because of Roman determination to avoid further imperial interference.

The next four elections, including that of Leo I the Great in 440, were undisputed. Leo was in Gaul on a diplomatic mission when he was elected, the first pope to be absent from Rome at his election. He was also the first to be buried in St. Peter's basilica. Simplicius, the second pope after Leo, was ill long before his death in 483 and had time to establish procedures for his successor's election. By then there no longer was a Western emperor, Romulus Augustulus having been deposed in 476, and the Germanic general Odoacer held power in Italy. Simplicius decreed that Odoacer's minister in Rome, a Roman nobleman, approve the choice of his successor. Apparently he wanted to ensure that a Roman had final say on the next pope and not Odoacer, who was an outsider and an Arian Christian. The choice fell on a Roman aristocrat, Felix III (or II, if the earlier Felix is deemed an antipope), the first bishop of Rome who clearly came from the patrician class. He was a priest's son and a widower with three children. Felix's successor was Gelasius I (492–96), who was consecrated on the day Felix died. His place in papal history lies in being the first to call himself "Vicar of Christ" and in his "two swords" theory, which proclaimed that there are two powers in the world, the church, centered in the pope, and the state, centered in the emperor, with the church the superior.

Gelasius's hard line toward the Greek bishops and the Eastern emperor resulted in a backlash with the election of Anastasius II in 496, who sought reconciliation with the East. Anastasius's policy had strong opponents in the Roman clergy, so his death two years later led to another disputed election. The anti-Anastasius faction chose Symmachus, a Sardinian whose parents were pagans, while the pro-Anastasius group elected Lawrence, from an old Roman family. Symmachus controlled the district around St. Peter's, and Lawrence held the Basilica of St. John Lateran. Unrest and violence erupted in Rome, and members of both sides were killed. The factions appealed to Theodoric, the Ostrogoth king of Italy, although he was an Arian. Both factions gave money to royal councilors, if not to the king himself, to influence the outcome. This is the first known case of bribery, later defined as the sin of simony, in a papal election. Theodoric gave the office to Symmachus because he had more support in Rome and had been consecrated first, although Lawrence's followers challenged his control of the Roman church for almost a decade. Symmachus was eager to prevent a similar problem in the future. He issued a decree allowing the reigning bishop to designate his successor while prohibiting any electioneering before and after a pope's death, especially by members of the Roman Senate, who opposed him. For perhaps a half-century the laity of the Roman community was removed from participation in the papal election.

In 514 the clergy selected Hormisdas, who had helped arrange loans for Symmachus to argue his case before Theodoric. He had been married before his election, and his son would become pope in 536. His election and that of John I in 523 were made without disruption. John's pro-Byzantine policy led Theodoric to order him to come to Ravenna, where he died in 526 in prison, leading to his being honored as martyr and saint. The Ostrogoth king then intervened in the papal election, imposing Felix IV on the Romans. Felix was eager to ensure the control of his faction over the papacy. Approaching death in 530, he summoned the clergy of Rome to his sickbed and gave his *pallium,* a cloth symbolic of episcopal office, to Boniface, an archdeacon who had Germanic ancestors. Felix decreed excommunication for anyone who refused to accept his choice for successor. The Roman Senate objected to the absence of an election and denounced Felix's acts. It reaffirmed a decree from Anastasius II prohibiting a pope from designating his successor.

When the pope died a few days later, a majority of Roman clergy chose Dioscorus, while a minority supported Boniface. Dioscorus's death a month later halted the mounting schism, and Boniface II successfully became pope.

The clergymen supporting Dioscorus were obliged to acknowledge that they had been in schism before they were allowed to resume their offices. Boniface attempted to name his successor, but the resulting uproar forced him to back down. The election that followed his death in late 532 was highly disputed and filled with charges of bribery and coercion. It took two months to choose an ordinary priest as a compromise. His name was Mercurius, and he became John to avoid polluting the papacy with the name of a pagan god, being the first pope to change his name after election. It took five centuries for the practice to become standard. The Ostrogoth king Athalaric pressed John II into approving decrees barring any private agreements to elect a pope and limiting drastically the amount of money that could be spent during a papal election. John died two years later, and after Agaptius I's brief reign, Athalaric pushed through the election of Pope Hormisdas's son Silverius.

By then the Byzantine reconquest of Italy was well under way, and Emperor Justinian set up a rival to Pope Silverius in Vigilius, who had been serving as papal legate in Constantinople. Byzantine forces seized Silverius and forced him to abdicate in 537 in favor of Vigilius, who had been Boniface II's choice for his successor. Silverius died a month later. Late in Vigilius's eighteen-year reign, he incurred Justinian's wrath and died in exile in 555. The emperor then appointed Pelagius I, only later holding a sham election. The Romans accepted Pelagius most begrudgingly, and he accomplished little during his five years as pope. The emperor allowed an election for his successor but insisted on approving the selection before John III could be consecrated. That remained the practice for a century.

Information is limited on the election of the next four popes, including Gregory I the Great's in 590, despite vast source material on his life and reign. It is known that the clergy and the senators, who now represented the Roman people, chose him unanimously and that he wrote to the Byzantine emperor begging him to refuse consent because he felt overwhelmed by the tasks the new pope would face. The emperor did not, and Gregory went on to become one of the great popes. As a member of a patrician Roman family and great-great-grandson of Felix II, he had the training and family tradition to become a strong administrator. Gregory was also the first monk to be elected pope and favored monks for offices in the Roman church.

As has often happened in papal elections, a powerful pope with a clear direction of policy was succeeded by one with a different approach to church affairs. In 604 the Roman clergy elected Sabinian, a deacon, largely to counter Gregory's favoritism to monks, but he lost popular support because of a Lombard siege, floods, and famine. He was accused of selling grain when previously it was given out free during times of famine. At his death in 606, a dispute between pro- and anti-Gregorians led to a year's delay of in consecrating Boniface III, a supporter of Gregory. His reign of less than a year was marked by a decree denouncing bribery in the papal election and forbidding any discussion of papal candidates until three days after the funeral of the dead pope. Then there would be a meeting of the clergy and the "sons of the Church," apparently the nobles, to elect his successor. Each was to vote according to his conscience.

The next four elections occurred quickly after each pope's death, indicating no serious factionalism and quick imperial approval. When Severinus was elected in 640, however, he waited twenty months for the Byzantine emperor's approval because of a dispute between the Latin and Greek churches. Thus, his reign as given on the official list of popes lasted only two months, since he died soon after imperial approval arrived. Martin I refused to wait for the emperor to accept him and arranged for his consecration within days of his election. This and Martin's staunch opposition to the Greek position in doctrine enraged Constans II, who sent troops in 653 to take him to Constantinople for trial for treason. He was found guilty, but his death penalty was commuted to flogging and exile. Meanwhile, the Roman church proceeded to elect Eugene I, an elderly priest. The official list of popes has an overlap of over a year between the time Eugene was elected and Martin died in exile, during which time there were two legitimate popes. Martin was deemed a martyr for the faith, the last pope so designated. The next seven popes were more accommodating to the emperor, and imperial confirmation was made without delay. But Benedict II had to wait a year for imperial approval in 684. He convinced the emperor to allow the Byzantine governor of Italy at Ravenna to confirm a new pope. That then became the practice.

By this time elections were being held in the basilica of St. John Lateran, which has served as the cathedral for the bishop of Rome since the fourth century. When a disputed election occurred, the faction that held the Lateran was far more likely to win. The Lateran complex included a papal palace as well as the basilica, and it was the domain of the clergy, which suggests that they dominated the elections, but there is little evidence on who attended or what the electoral procedure was. The basilica was too small to hold all the Romans, but the continued use of the phrase "with the whole people" indicates that some lay participation still occurred.[8] Since St. Peter's was outside of Rome's walls, it lacked an adequate residence until it was enclosed in walls after an Arab raid in 846.

When John V, a native of Antioch, died in 686, a new factor, the Roman army, made up of Roman aristocrats, became involved in papal elections. The clergy supported one candidate; the soldiers, another. The soldiers seized control of the Lateran and forced the clergy out, but that did not end the stalemate. It ended only when Conon, an elderly priest and son of a Byzantine general, emerged as a compromise choice. The army involved itself again after Conon's death in 687 when factionalism in the clergy prevented a decision by the clergy. Both factions seized control of parts of the Lateran. After a month's stalemate the soldiers intervened and arranged for the election of Sergius I as a compromise. A faction refused to concede, and the Lateran had to be taken by force in order to consecrate Sergius, who reigned until 701.

The army was quicker off the mark for the next two elections, insuring that its candidates were selected without dispute or violence. In 708 it was off to war, and the Roman clergy, in the sort of fit of piety that occasionally has disrupted the papacy, elected Sisinnius, a deeply devout man. Perhaps fortunately for the Church, his was one of the shortest reigns in history, only three weeks. Constantine, his successor, was last in a series of seven popes who were Easterners, although all had been active in the Roman church long before their elections. Gregory II, from old Roman aristocracy, followed him in 715. Gregory was true to his background by being an excellent administrator during his sixteen-year reign. During his funeral in 731, the Roman crowd hoisted up Gregory III and acclaimed him as pope. His successor, Zacharias, was the last pope to announce his election, in 741, to the Byzantine authorities.

Stephen II was elected unanimously in 752.[9] He crossed the Alps in 753, the first pope to do so, to beg the Frankish ruler Pepin for protection against the Lombards. Not only did Pepin respond favorably to the appeal, he also recognized papal sovereignty over Rome and lands largely to the northeast of the city. This Donation of Pepin did not create the Papal States, since the pope already had extensive political authority in the region, but it made manifest that the pope was a secular ruler. That in turn meant that the Roman nobility would become far more interested in holding papal office or at least controlling who would be elected. That fact was demonstrated in the disputed election that followed Stephen's death in 757 in which his brother Paul I was chosen. Their pro-Frank policy had opponents still loyal to Constantinople, while Paul was criticized for paying little attention to the interests of Roman nobles.

The election dispute that followed Paul's death in 767 had a major impact on the system of choosing the pope. Toto, a nobleman, gathered an armed force and seized the Lateran, where his brother Constantine was proclaimed pope, while he took the title of Duke of Rome. Constantine was a layman, and three bishops were coerced into ordaining him priest and bishop. Although he had broad support in Rome, he also had opponents who called on the Lombards for aid. Street fighting broke out, and Toto was killed. The Lombards sought to establish their own choice as pope but found no support. A year after Constantine's election, the political situation had changed enough that a more regular election held in the Forum by clergy and laymen succeeded in electing a priest as Stephen III. Constantine was dragged from the Lateran, had his eyes gouged out, and was imprisoned in a monastery. Stephen III held a synod of the province of Italy in 769 that voided Constantine's election and all his acts except baptisms. More important, it decreed that the entire Roman clergy had the right to elect the pope but only the cardinal-priests and cardinal-deacons were eligible to be chosen. That was the first use of the term cardinal to refer to the priests of the major parishes of the city and the seven deacons. The cardinal-bishops of the nearby cities were excluded, probably because several had supported Toto. Only after the clergy had made their choice would the laity salute him as their lord. Although the Roman laity soon regained its role and kept it until

1059, the principle that the papal electors should be a small number of clergymen associated with the Roman church has remained in place to the present, although the number of eligible clerics would rise over the centuries.

Three years after the synod issued its rules, the first election under them took place without incident. Hadrian I, a nobleman who had risen in the ranks of the Roman clergy, was chosen unanimously. His reign of almost twenty-four years was the longest yet and the fifth longest in history. He filled the Roman clergy with fellow nobles. It was a peaceful period in Rome during which the papacy forged a close relationship with Charlemagne. Leo III, Hadrian's successor, needed that relationship to hold on to his office. He was not noble but was also elected without opposition, because his supporters acted the same day Hadrian died to get him chosen. He soon made powerful enemies who drove him from Rome in 799. He fled to Charlemagne's court, where he received a military escort to return to Rome. A year later Leo crowned Charlemagne as Emperor of the Romans when he came to Rome. The next two popes were also elected unanimously, but the election in 824 was filled with acrimony. Pro- and anti-Frankish factions, largely identified as the clergy and the nobles, promoted rival candidates. Four months of deadlock required the intervention of Louis the Pious, Charlemagne's son, to secure the election for Eugene II. Soon a new constitution returned the process to what it had been before 769. It reestablished participation by the laity, specifically the nobility, and the Roman nobles dominated papal elections for the next two centuries. It also required that the new pope swear an oath of loyalty to the Frankish ruler. In 827 a nobleman who was a cardinal priest became Gregory IV. He was not consecrated for six months, until a report on the election and his oath reached Louis and were accepted.

Lay participation in papal elections became a problem upon Gregory's death in 844, when the clergy elected one candidate, and the nobles, another. Emperor Lothair intervened in support of the nobles' candidate, Sergius II, who reaffirmed the taking of the oath to the Frankish emperor. Three years later the threat of an Arab attack on Rome persuaded the Romans to elect Leo IV even before Sergius's funeral. The Romans carried him in triumph to the Lateran, and he was consecrated without approval from the emperor. The breakup of the Carolingian Empire was making it difficult for the emperor to enforce his will on the papacy. In 855 Lothair II failed to establish his choice as pope in place of the man elected by the Romans, Benedict III. He became pope only after the Romans's first choice refused the office, the first time that is known to have happened. Lothair was present, however, when Nicholas I was consecrated after a unanimous election three years later. The new pope quickly issued a decree prohibiting anyone outside of the Roman community to be involved in a papal election. When Hadrian II, a relative to Sergius II, was elected in 867, the requirement of informing the Frankish emperor was ignored without consequence.

The turbulence of Hadrian's reign did not match that of his successor, John VIII, who was assassinated by a cleric in his entourage after a ten-year reign. His murder began what most historians would agree was the worst period in the history of the papacy. After a period of long reigns, popes now came and went in rapid succession. As many as twelve popes were assassinated, several after having been deposed first. Three more were deposed without being killed, and two abdicated. The era may have also seen the election of a woman, "Pope Joan."[10] An important innovation in papal elections occurred in 882 when Marinus I was chosen despite being a bishop of another diocese. That was contrary to a tradition codified in canon law at Nicaea that no bishop could move to another see. The violation had no consequence for Marinus, but it did for Pope Formosus, also a bishop, when he was elected in 891. Soon after his death in 896, his successor, Stephen VI, ordered the corpse taken out of its grave and put on trial for violating the law. Formosus was denounced as an antipope, and all of his acts were nullified. Since he had ordained Stephen bishop of a nearby see, the trial probably was a way for Stephen to avoid the charge that he himself had violated canon law. (In 1059 church law on that point was changed, and Formosus was returned to the official list of popes.) Stephen's act stirred up violent opposition, and he was deposed and then murdered. In less than three months in 897–98 there were three popes. One was deposed for not being hostile enough to Stephen's supporters, and the next was murdered for being too hostile to them, both about a month after their elections. John IX, the third pope, managed to die a natural death after a two-year reign. His successor Benedict IV may have been murdered in 903, and the same fate is certain for Leo V the next year.

The responsibility for Leo's death probably belonged to his successor, Sergius III. Sergius had been elected by a faction in Rome in 897 but had lost out. In 904 he gathered a force and marched on the city to oust Leo and have himself elected. He always dated his reign as beginning in 897. Sergius was allied with a powerful Roman nobleman, Theophylact, with whose daughter Marozia he had a son. Four of the next five popes were Theophylact's creatures. In 931 the power of the family reached its height when Sergius and Marozia's son became Pope John XI. A new power, King Alberic II of Provence, appeared on the scene four years later and deposed John. The next four popes owed their office to his control over Rome. Alberic's control over the papacy peaked in 955 when he forced the Romans to accept his son as the next pope. Octavian, the son, was duly elected at age eighteen, making him the youngest pope. Changing his name, he became John XII. Despite his youth, his reign was rather short. He died in 964, allegedly worn out by debauchery and excess.

The one noteworthy event of John's reign was the coronation of Otto I, king of Germany, as Holy Roman Emperor in 962. Thus began the identification between German king and Holy Roman emperor that would last for 850 years. Otto took an active interest in the papacy and regarded himself as its patron and overseer. Tales of John XII's immorality had persuaded him to hold a synod at Rome in 963 that deposed John and elected a prominent Roman layman as Leo VIII. Prior attempts to depose popes for personal immorality had ended in the decision that the pope could not be judged by anyone but God. It is therefore questionable whether Leo's election was valid. Those Romans who believed it was not valid ousted Leo and restored John once Otto left Rome. John died soon after, but the Romans, still refusing to accept Leo, proceeded to elect Benedict V. Otto returned to Rome in force and arrested Benedict as a usurper. A synod declared him an antipope, and Leo himself broke Benedict's papal staff over his head as he lay prostrate. Benedict was exiled from Rome, while Leo died in 965. Otto then secured the election of a bishop from Umbria as John XIII, ignoring the prohibition against moving a bishop from one see to another.

Otto was in Germany at John XIII's death in 972, but he had enough influence in Rome to gain the election of another client as Benedict VI. The

Roman nobles rejected Benedict, and when Otto died two years later, they murdered him and elected a successor, Boniface VII. His legitimacy remains uncertain: The next Pope Boniface was numbered VIII, but Boniface VII was stricken from the official list of popes in 1904. Otto II soon arrived to drive Boniface from Rome and arrange for Benedict VII's election. Boniface did not go away, however. He led an armed takeover of Rome in 980 that exiled Benedict. Otto II returned in force and, not bothering with an election, installed the bishop of Pavia as John XIV. He changed his name from Peter, not wishing to be so presumptuous as to be Peter II. Soon after, Otto II died, and Boniface VII returned to Rome to reclaim the papacy with the support of his noble allies. John was deposed and imprisoned, dying either of starvation or poison in 984. A year later Boniface probably was murdered, and the Roman nobles chose John XV as pope. He managed to stay in office until his natural death in 996, although Emperor Otto III was then on his way to Rome to depose him. Otto arranged for the election of his chaplain Bruno, who was the first German pope. He changed his name to Gregory V out of respect, he said, for Gregory I, but probably a more important motive for the change was that his original name made his non-Roman origins obvious. Taking a new name now became routine for a new pope.

Gregory V, as the non-Roman creature of the German emperor, was forced to flee Rome after Otto returned to Germany. The Romans elect their own man as John XVI, who is regarded as an antipope, but his number remains in the list of popes. Otto restored Gregory to power and was still in Rome upon Gregory's death in 999. Otto then saw to the election of his tutor Gerbert of Reims, who became Sylvester II in honor of Sylvester I, the model of cooperation between pope and emperor. They had little time to work together, as Otto died in 1002, soon followed by Sylvester. The Roman nobles chose the next three popes. A rival noble faction chose one of its own as Benedict VIII. He was followed by his brother as John XIX and his son as Benedict IX.

Few papal reigns have been as confused as Benedict IX's, who served as pope in three periods from 1032 to 1048. Sufficient to say that when Henry III arrived at Rome from Germany in 1046 seeking coronation as emperor, he found three rival popes. He deposed all three and installed his own choice as Clement II. Three years later, as a source from the era states, "Leo IX was enthroned in the Roman see by the emperor and his nobles,"[11] but the Romans acclaimed him as pope when he arrived at Rome in 1049. Over the next ten years Henry III appointed three more popes, all Germans, without bothering with the formality of election. Henry and his popes were eager to reform the papacy, but they had no interest in changing the system of papal selection. Had Henry lived a full life instead of dying young in 1056 and had his strong-willed son become emperor at a mature age, they probably would have succeeded in eliminating any semblance of election and making the office entirely appointive. As it was, Henry IV became emperor as a child and lost control of the papacy, which cleared the way for church reformers to draw up the election rules of 1059 that drastically transformed the nature of papal elections.

CHAPTER II

Election by the College of Cardinals, 1059–1274

By 1059 the papacy had become a powerful institution controlling not only the religious life of Western Europeans but also extensive revenues and a large state in central Italy. Naturally, the question of who would control that much power and wealth was a major concern for many, not least the Holy Roman Emperor. By then, however, the Catholic clergy had developed a sense of being a separate class with its own interests and goals. The issue of who would choose the pope became a point of great contention between emperor and churchmen, and thereby between laity and clergy. There was much at stake, and the issue's resolution went a long way toward determining the nature of medieval society.

After the conversion of Constantine, Christians largely had accepted the concept of caesaropapism: The emperor as the highest secular power had the right to direct the affairs of the Church. The Greek church and its patriarchs remained under the Byzantine emperor's control to the fall of the Byzantine Empire, while the Latin church was forced to redirect its focus from the Roman emperor to in turn the Byzantine, the Carolingian, and the Holy Roman emperors. Between each period of domination by successive empires, the papacy was largely free of any secular power, and the Roman clergy, and laity to a lesser extent, selected the pope without outside interference. For some churchmen, that was the proper state of affairs, and the fact that they gained a taste of acting on their own was a powerful incentive for their drive to be free of secular control. Their model was the great monastery of Cluny, which from its founding in 910 in eastern France was independent of secular authorities. By placing itself directly under the pope, Cluny was free from secular interference while it had nothing to fear from the distant and weak papacy of the era. The amazing success of the Cluniac monks in spreading their houses across Europe, which answered to the abbot at Cluny and not to any local authority, whether clerical or lay, spread the ideology of an independent clergy.

The popes, under the thumb of the Roman nobility in the early eleventh century, were slow to reflect that new clericalism. Holy Roman Emperor Henry III brought reform to the papacy by appointing popes influenced by the Cluniac movement, especially Leo IX, his cousin. When Henry died in 1056, leaving a six-year-old son as his successor, the reformers were ready to apply their principle of an independent clergy to the papacy. A year after Henry's death, his handpicked pope, Victor II, died, and the Roman clergy, acting quickly, held a relatively free election for the first time in a century. They chose the abbot of the monastery of Monte Cassino, a native of Lorraine, as Stephen IX. He informed the Imperial court in Germany only after he was consecrated. His short reign is noteworthy for the appointment of three great churchmen—Peter Damian, Humbert, and Hilderbrand—to positions in the Roman church, thereby making them eligible to vote and be elected in the papal elections. Hilderbrand had been sent to the Imperial court to inform it of the new pope's election, and in 1058 a dying Stephen required an oath from the Romans that they would not elect a pope until he returned. Delaying the election suggests that Stephen was sure that the Roman nobility could not control it. It almost was a mistake, as the Roman nobles with some priests assembled and elected a local bishop who took the name of Benedict X. Damian, as cardinal-bishop of Ostia, refused to consecrate him and, with the other cardinal-bishops, fled Rome for Siena, where they elected Bishop Gerard of Florence, a Burgundian by birth, as Nicholas II. In January 1059 in the company of troops supplied by the duke of Lorraine, Nicholas entered Rome and was consecrated in the Lateran. His consecration added elements taken from the Byzantine emperor's coronation to the traditional service. Henceforth the term papal coronation is appropriate. Benedict was condemned as an antipope.

Nicholas's place in history comes largely from the synod he held in April 1059 to deal with the method of the election of the pope. At least seventy-nine bishops attended. One decree that came from it condemned simony, defined as buying a church office. While intended as a reform measure to ensure a properly selected clergy, it also gave the papacy a powerful tool to use against recalcitrant prelates, who often did buy their offices, should they offend the pope. It proved especially useful to the papacy when investiture, in which a secular ruler invested a new bishop with his symbols of office, was condemned as simony in 1078. It was, however, a two-edged sword, as it could also be used, as it often was in centuries to come, to try to depose popes for buying their offices. Although no pope ever was removed, accusations of simony made life difficult for several later ones.

Turning his attention to the papal election, Nicholas first made reference to the disorders that followed the death of Stephen IX, which, he said, were so severe that the "Chief Fisherman" came close to shipwreck.[1] He proposed a set of rules lest the same evils arise in the future. The election must not be held before the dead pope has been properly buried. The cardinal-bishops are given the responsibility first to certify that an election must be held and then identify a candidate from the Roman clergymen; if a suitable man cannot be found among them, one can be chosen from elsewhere. The other cardinals give their consent to the candidate identified by the cardinal-bishops, followed by the remaining clergy and the laity. The election should take place in Rome, but "if the perversity of depraved and wicked men shall so prevail that a pure, genuine, and free election can not be held in Rome," then the cardinal-bishops along with the clergy and laity, even if few in number, can meet wherever it seems to them most suitable. Once the new pope has been elected, he has the authority at once to rule the Church and control its resources, even if he cannot be enthroned in the Lateran basilica because of war or other problems. This suggests that immediate consecration is proper, which in turn implies that Imperial confirmation is unnecessary. If the new pope is not a bishop, his power to ordain other bishops must wait until his own ordination as bishop. The decree goes on to threaten excommunication against those who would violate these rules, denouncing them as Antichrists and destroyers of Christianity.

The decree does make an ambiguous reference to the honor and reverence due to Henry IV and his successors as emperor. Some scholars argue that it affirmed the old tradition that the emperor had the right to approve a new pope although certainly not the power to appoint one; others, that it reduced if not eliminated the emperor's role in approving a new pope. A second version of the decree, however, recognized the emperor's special role. It is not clear whether this alternative version was written to appease Henry or was an Imperial forgery done during the Investiture Controversy. It is vague in what power it concedes to the emperor. It adds him to those who take the lead in electing the pope, and the clause allowing the papal election to occur away from Rome is changed to give him a role in deciding where it would take place. Future emperors would cite it to assert their rights in choosing or at least confirming the new pope, but the papacy successfully ignored it, although new popes for another century informed the emperors in a special way of their election.

Since Nicholas had made many decisions before his eventual enthronement in the Lateran, he and other reformers were eager to have them deemed valid. This decree, therefore, was first of all a retroactive validation of Nicholas's election, which had three defects under the rules in effect when he was selected, although they often had been violated before. His election took place outside of Rome; he was not a member of the Roman clergy; and he was already a bishop of another see. In the decree Nicholas changed several long-standing traditions in respect to the papal election with little in the way of justifying them in the form of citations from Scripture and canon law, although he did cite Pope Leo I's letter of 458 requiring the clergy to choose a bishop.

To be sure, the new decree was similar to Stephen III's of 769 limiting papal electors to the Roman clergy and candidates to the cardinal-bishops, but it made several changes that had a huge impact in church history. One was making non-Romans eligible to be elected pope, at least under special circumstances. It signaled that the papacy was ready to take seriously its claim to have universal authority over the Church by going beyond the

Roman clergy for its incumbents. While the cardinals of 1059 were clergy-men living in Rome who were not always Roman by birth, the pope's universal authority would soon require that some represent the Catholic nations more fully by living distant from Rome. The option of electing the pope away from Rome, which occurred in most elections over the next 250 years, made possible the Avignon papacy. Perhaps the greatest impact of the decree of 1059 was that it widened the divide between clergy and laity. Certainly that gap had been spreading for centuries, but removing the Roman laity from a meaningful role in electing the pope and drastically reducing the emperor's influence was intended to give the clergy the only effective voice in the election. Lay influence on future papal elections would be present, but it could be wielded only by bribing or coercing the cardinals.

The Imperial court seems to have taken little immediate notice of Nicholas's election decree, but it vehemently denounced him when he made peace with the Normans in southern Italy. In the previous half-century adventurers from Normandy had successfully seized control of southern Italy from the Byzantines and the Muslims, and both the Holy Roman emperor and the papacy had opposed them. Leo IX had personally led an army against them only to be defeated. Now, much to Imperial rage, the pope recognized their right to rule the region and allied himself with them. The Normans promised to provide military force to protect the pope and guarantee the freedom of papal elections in the manner established by Nicholas. A synod of German bishops declared his acts invalid, including his election decree.

Two rival popes appeared after Nicholas's death in July 1061. Delegates from the Roman nobility and those clergymen opposed to him went to Henry IV asking him to appoint a true pope. He chose the bishop of Parma, who took the name Honorius II. The reformers, made up of a majority of the cardinals and the Roman clergy, elected the bishop of Lucca, who became Alexander II. He was enthroned in the Lateran the next day. Normans came to Rome to protect him against Honorius, but Honorius's force of Roman nobles and Imperial troops succeeded in driving him from Rome. The duke of Lorraine persuaded both men to attend a synod at Mantua in 1064 to determine the true pope. Honorius expected to preside, and when he was denied, he withdrew. Alexander presided by default, and after swearing that he had not gained the papacy by simony, he was acclaimed rightful pope. Honorius retreated to Parma but never renounced his claim to be pope. Alexander meanwhile went from success to success, persuading synods of bishops across Europe to denounce simony and clerical marriage.

Upon Alexander's death in 1073, the Romans moved swiftly to acclaim Hilderbrand as Pope Gregory VII before Henry IV could act to control the election. The new pope was not a native of Rome but had vast influence in the Roman church as archdeacon. The official protocol of the election stated that a huge assembly met in St. Peter's basilica. "We—the cardinals, clerks, acolytes, subdeacons, deacons and priests of the holy Roman catholic and apostolic church, in the presence and with the consent of the venerable bishops and abbots, clerks, and monks, and with the acclamation of many crowds of both sexes and various orders—elected as our pastor and supreme pontiff . . . the archdeacon Hilderbrand."[2] Gregory himself referred to the great tumult of people who fell on him like madmen and forced him with violent hands onto the apostolic throne. Thus Gregory, who had so much to do with the election decree of 1059 (having written it, according to his foes), was made pope contrary to its provisions. His era, however, was not far removed from the time when popes were elected by public acclamation, and few disputed the idea that God had inspired the masses to choose him.

Among those who did deny the validity of Gregory's election was Henry IV. He may have accepted him at the moment of hearing of his election, but within a year Gregory's prohibition of royal investiture of bishops led to a showdown between them. When the pope condemned the king for investing the royal candidate for the major see of Milan, Henry in 1076 assembled the German and Italian bishops loyal to him, who denounced Gregory's election as illegal according to the decree of 1059. They argued that his election was invalid because it occurred before his predecessor had been buried, although Gregory's supporters maintained that he was devoutly attending the funeral service when the popular uproar to elect him began. Gregory was also accused of violating an oath he allegedly took to Emperor Henry III never to accept the tiara.[3] The pope responded by excommunicating Henry and his bishops and absolving the German nobles of their oaths to him. As is well known, Henry saved his throne by going to Italy and kneeling in snow for three days before the castle at Canossa, begging and receiving forgiveness from Gregory. The struggle was far from over, however. In 1080, Henry called another synod, which included thirteen cardinals fearful of Gregory's rigidity, to depose the pope. The synod denounced him for using money stolen from the papal treasury to buy popular support for his election and assembling an armed force at the time of Nicholas's funeral to coerce the people into electing him. It proceeded to elect its own pope, Clement III. Gregory for his part reexcommunicated Henry and deposed him in favor of a German rival. Henry got the upper hand when he occupied Rome in 1084, forcing Gregory to flee to the Normans in southern Italy. There he died the next year.

Gregory had overreached in his efforts to depose Henry, but his powerful statements in support of the pope's right to do so were major weapons in the arsenal of the papal monarchy of the next centuries. By his words at least, he can be regarded as the founder of that papal monarchy. Among his contributions to the growth of papal supremacy was his decree that only the bishop of Rome could be called pope, ending the by-then rare use of the title by other bishops. Yet Gregory left the papacy in shambles. He violated a long-standing rule by naming three men from whom he wanted the next pope to be chosen. The men he nominated were non-Romans, which caused resentment in Rome. It took a year for the reform party, reduced by defections to Clement III, to hold an election.

The controversy over the electoral rules of 1059 during this period is made clear in the collection of papal canons made in 1086 by Deusdedit, a prominent cardinal-priest. He was unswervingly hostile to the new rules and did not accept giving any role in the election to the cardinal-bishops or the secular rulers. The election must take place in Rome, and the man chosen must be Roman except under the most exceptional circumstances. The clergy of Rome must elect the pope and have him acclaimed by the Roman people. The bishops as a group have no right to participate in the election, nor can a dying pope designate his successor, although Deusdedit would allow him to suggest some names.[4]

Few reformers were monks, but they regarded monks as free from the simony and sexual license that permeated the secular clergy, especially the bishops. They saw abbots as men of pure faith, with the administrative experience to govern the Church. A majority of the popes in the century after 1086 were monks, beginning with Abbot Desiderius of Monte Cassino. The reformers meeting in Rome chose him as pope, but he at first refused the office. He was not among those Gregory had designated, and the most loyal of the dead pope's supporters opposed his election. In 1087 the electoral process was resumed in southern Italy under Norman protection, and this time the abbot accepted the papacy, taking the name Victor III. He hoped thereby to signal his goodwill to Henry IV, since Henry's father had appointed Victor II. The Romans, not Henry, prevented Victor setting foot in Rome for six months, because he was an outsider. Four days after arriving in Rome, he was forced to flee again. He renounced his election and went back to Monte Cassino. Late in 1087 his supporters gained control of the Lateran, and he returned to Rome to be crowned. He died two months later at Monte Cassino, after Clement III's supporters had won back Rome.

The reformers again met in southern Italy to hold an election. A larger group than that which had elected Victor chose the French-born cardinal-bishop of Ostia as Urban II.[5] This appears to be the first papal election in which Nicholas II's rules were followed fairly closely, even if Victor had designated Urban as his successor. The four cardinal-bishops present consecrated him pope the same day. Because he continued Gregory's policies almost entirely, he could not reside in Rome until 1093, when Henry IV's rebellious son seized the city from his father's allies and invited him to return. Urban's reign is noteworthy for the First Crusade, but he died in 1099 without hearing that the Crusaders had taken Jerusalem two weeks before. Although Clement III was still alive, the election of Urban's successor, held in Rome, followed the 1059 decree. The new pope, Paschal II, was a cardinal-priest of Rome. His nineteen-year reign, the longest in 300 years, continued the Investiture Controversy not only with the emperor but also the kings of France and England. The Roman opposition to Paschal forced him out and established three antipopes, and he returned only a few days before he died in 1118.

The election of Paschal's successor was the first one conducted under the election decree of 1059 for which a detailed account exists, thanks to a biography of the pope written a decade later. It indicates that a total of forty-nine cardinals—four cardinal-bishops, twenty-seven cardinal-priests, and eighteen cardinal-deacons—along with the rest of the Roman clergy and many prominent laymen, were present when a cardinal-deacon was unanimously elected as Gelasius II. The election did not take place at the Lateran but in a monastery because of powerful opposition to the dead pope led by the Frangipani family. As soon as word of the election got out, the Frangipani imprisoned Gelasius before he could be ordained a priest and bishop. He agreed to leave Rome and was ordained at Gaeta, near Naples. A failed attempt to return to Rome forced Gelasius to flee to France, where he died at Cluny in early 1119.

Four cardinal-bishops had accompanied Gelasius to France, and at his death they proceeded to hold an election there because Rome was under the control of an antipope erected by Henry V, the new emperor. Gelasius had clearly indicated his choice for pope on his deathbed, and the cardinals agreed, electing the archbishop of Vienne in France as Callistus II. Letters were immediately sent to Rome seeking the assent of the cardinal-priests and cardinal-deacons, but Callistus was so confident of the validity of his election that he arranged for his coronation in the cathedral of Vienne before word came to him of their acceptance. This election is important in that both election and coronation took place far from Rome and for its clear indication that a small number of cardinals could elect a pope under special circumstances.

The rights and powers of the cardinals were beginning to coalesce into what would soon be known as the College of Cardinals. The first use of the term is dated to 1148. The College consisted of the traditional seven cardinal-bishops; twenty-eight cardinal-priests, representing the major churches in Rome; and eighteen cardinal-deacons. In the decades to come the number of cardinals steadily declined, until numbers in the teens became routine. For a century non-Romans had been appointed as cardinal-bishops, but the cardinal-priests and -deacons had remained a preserve of the Roman clergy. That situation was one reason why the Romans had made an effort to eliminate the special role of the cardinal-bishops found in the decree of 1059. By 1119 non-Romans were appearing in the other two ranks as well.

As papal power increased, so did the influence of the cardinals as a body. They became the popes' closest advisors and were soon hearing legal cases, witnessing papal bulls, and being sent as legates across Europe. For a time the term "Senate of the Roman Church" was used for the body of papal advisors and administrators, but the one that became permanent was *curia,* the Latin word for court, reflecting the ever-increasing monarchical nature of the papacy. By 1200 papal meetings with the cardinals were known as consistories. Usually they were very frequent, almost daily under some popes.

The sense among the cardinals by 1119 that they were a body with common interests explains why there was little opposition to Callistus's election by a small part of the group. It was only fourteen months later that the pope entered Rome, where he was greeted with enthusiasm despite not being a Roman nor having had any previous ties to the city. Soon after arriving in Rome, Callistus, a reformer but not heavily burdened by the Gregorian legacy, began negotiations with Emperor Henry V. They culminated in the Concordat of Worms of 1122, which settled the Investiture Controversy largely in the pope's favor. Two years later Callistus died in Rome.

Any hope that papal elections might become routine disappeared at Callistus's death. Two families, the Pierleoni, allied with the Normans, and the Frangipani, allied with the emperor, had emerged as the leaders of major political factions in Rome. When the cardinals and bishops met the required three days after the pope's death, they quickly elected a cardinal-priest supported by the Pierleoni. He took the name Celestine III and was undergoing the ceremony of consecration when the Frangipani burst into the church at the head of an armed force and demanded that their candidate be made pope. Violence ensued, and Celestine was wounded. He quickly renounced his election, and the cardinal-bishop of Ostia was acclaimed as Pope Honorius II. The next day the cardinals reassembled to regularize the situation. Allegedly the Frangipani had bought off the senior cardinals with large bribes overnight, and they now supported Honorius, who first renounced his acclamation of the previous day and then was duly elected. The official list of popes does not include Celestine.

Honorious, a pious monk from Bologna, seems unlikely to have been the willing tool of Roman aristocrats such as the Frangipani. It is more likely that they were acting in concert with a cardinal-deacon, Haimeric, a Burgundian whom Callistus had named chancellor of the Roman church. Research has suggested that Haimeric was the leader of the cardinals whom Callistus had appointed, who were about a third of the total.[6] In order for Haimeric to get the pope he wanted, he had to ally himself with the Frangipani. If this scenario is correct, it marks the first example of a well-defined split in the College of Cardinals having an impact on the papal election. Haimeric apparently did not want to become pope himself but to control the election of the next pope, probably because he wanted one opposed to the Normans. This too is the first case of a powerful cardinal clearly serving as popemaker. Both phenomena have occurred frequently since 1124.

When Honorius fell seriously ill in early 1130, Haimeric secluded him in the monastery of Septasolis, a former Roman Imperial palace near the Coliseum, until he died. Haimeric then rushed the funeral so that the election would take place there as quickly as possible. Yet such manipulations failed to secure the easy enthroning of the man he wanted. The other cardinals insisted on organizing a committee of eight to establish the rules of the election and nominate a candidate who could be acclaimed by the whole Roman church. The committee was divided between those who had been cardinals before Callistus's election and those he named. The latter, Haimeric's supporters, chose the Cardinal of San Angelo, who took the name Innocent II. Support for him was limited, however; and a majority of the forty-one cardinals present elected Pietro Pierleone, a respected cardinal in his own right. He became Anacletus II. This division largely reflected the split between the pre- and the post-1119 cardinals. The latter included members from northern Italy and across the Alps. The Pierleoni supported their relative, and the Frangipani recognized Innocent. The Pierleoni controlled the Lateran, so their pope was crowned there, while Innocent was crowned in Haimeric's church.

Anacletus at first had the upper hand, winning broad support among the lesser clergy and people of Rome and controlling most strongholds of the city to the point that the Frangipani soon decided to recognize him. Innocent was forced to flee to France. Anacletus's campaign to be recognized across Catholic Europe was complicated, however, by the fact that his family had been Jews. As bankers to the papacy, they had become wealthy and turned Christian. They had been intimately associated with the papacy since 1070. The Romans clearly did not hold his Jewish ancestry against him, even if his rival's family was old Roman nobility. It became an issue when both popes' campaign to be accepted spread into northern Europe, where Innocent had more support. To what extent it was a factor in deciding for Anacletus or Innocent is impossible to say, but Innocent's supporters did draw attention to it. The schism also involved a debate over the proper role of cardinal-bishops in a papal election. The Roman clergy was not keen on giving them the dominant role found in the election decree of 1059, since most cardinal-bishops were not natives of the city. Innocent had the support of the cardinal-bishops, but only a few from the other two orders. Differing interpretations of the 1059 decree played a major role in the schism of 1130.

A century earlier, Anacletus's control of Rome would have been adequate to ensure him victory in the contest, but by 1130 it was not. The non-Italian Catholics largely favored Innocent not only because he was present in France but also because he had better contacts in the north, especially Bernard of Clairvaux, who was tireless in his efforts to get recognition for Innocent. Nonetheless, the schism remained a stalemate until Anacletus's death in 1138. Most of his supporters then recognized Innocent as pope, although a handful proceeded to elect an antipope, Victor, who within a year submitted to Innocent. In 1139 Innocent held the Second Lateran Council, where the first order of business was to condemn Anacletus and declare his acts and ordinations invalid. The official list of popes lists Innocent as pope and his rival as an antipope, but at the time objective observers would have been hard pressed to say who was the true pontiff.

Innocent II's triumph was largely a matter of outliving his rival, but it was also a victory for those who argued that all the cardinals had equal voting rights in a papal election. Deusdedit's argument restricting the electors to the Roman clergy was not heard of again, nor was the practice of having the cardinal-bishops serve as a nominating committee. Since all the cardinals now had equal electoral rights, the division between the cardinal-bishops representing the universal Church and the other two ranks standing for the clergy of Rome began to disappear. Succeeding popes began to appoint non-Romans and non-Italians to the other two ranks of cardinals, making the

College an international body and turning the Roman titles of the cardinal-priests and cardinal-deacons into honorary ones. The disputed election of 1130 also demonstrated that the election decree of 1059 did not solve the issues of the influence of politics and personalities in papal elections. Even as the papacy refined the electoral system, disputed elections and schisms continued to flourish.

The election after the death of Innocent II in 1143, however, was entirely consistent with the decree of 1059 and was not disputed. A cardinal-priest from Rome was chosen Celestine II, making the point that the Celestine of the disputed 1124 election was an antipope. Celestine's announcement of his election noted that he had been elected by the cardinals followed by the assent of the rest of the Roman clergy. His death five months later led to the election of Lucius II, another Roman cardinal-priest. Details of his election are largely unknown, but it was not controversial. The major event of his reign was the uprising in Rome against papal authority led by the brother of the late Anacletus II in 1144. Lucius fled the city and, gathering his own army, returned in force only to die from injuries he suffered when hit by stones during an attack on the city's walls. On the same day that Lucius was buried, the forty-four cardinals unanimously elected a Cistercian monk and pupil of Bernard of Clairvaux as Eugene III. Since the rebels controlled the city, he could not be crowned in the Lateran, but soon they agreed to accept him as pope and allowed him into Rome. He began construction of the palace alongside St. Peter's that became known as the Vatican. In 1146 the Roman rebellion took a more serious turn when Arnold of Bresica took over leadership of the Roman commune and injected a larger element of anti-clericalism into it. He denounced Eugene as "a man of blood." The pope fled from the city and died in exile in 1153.

His successor, Anastasius, a well-loved native of Rome, was accepted by the Roman people, who allowed him to be consecrated in the Lateran. His death a year later ended the possibility that the Roman commune and the papacy might reach an accommodation. His successor, Nicholas Breakspear, was elected unanimously by the cardinals but with no mention of the consent of the Romans. He has been the only English pope. As Hadrian IV, he took a hard line against the commune, and allied with Emperor Frederick I Barbarossa, he captured Arnold of Brescia and had him executed. Soon after, the two allies had their own falling out, and Hadrian died in 1159 a few days short of the deadline he had set for Barbarossa to submit or be excommunicated.

It was obvious that whenever the pope and the emperor were in conflict, there would be a disputed election. Hadrian sought to ensure a future anti-Imperial policy by appointing cardinals who opposed Barbarossa. Previous popes had left many vacancies in the College, which Hadrian had filled with anti-Imperialists. When the College and a portion of the clergy met in St. Peter's for the semipublic election, it agreed that some cardinals "should be deputed to listen to the opinions of others and should diligently investigate and record them accurately. If it was God's will that they could agree on any of the brethren, he should be elected with good will; if not, a man from outside [the College] should be considered and if they could agree on him, well and good."[7] Hadrian's work was successful in that a majority declared for his chancellor, Orlando Bandinelli, who took the name Alexander III. The five Imperial cardinals rejected the choice and selected Cardinal-bishop Ottaviani as Victor IV (the second antipope with that name). Violence erupted when Victor tried to seize the papal mantle that had been placed on Alexander. When he failed to get it, he produced a second mantle, which, however, he put on upside down and backward. His opponents proclaimed that his fumble was proof of satanic influence on him. This election was the last one in which the traditional phrase "with the assent of all the people of Rome" is known to have been used. Neither faction could crown its pope in the Lateran. The emperor claimed his right to mediate a disputed election and held a synod at Pavia in early 1160 with about fifty bishops present. Predictably it declared for Victor. Victor's supporters gained control of Rome, and Alexander fled to France. Victor's death in 1164 undercut the schism but did not end it, as two cardinals and some bishops elected a successor. Most Catholic rulers supported Alexander, and Frederick I was forced to accept him after the Imperial army suffered a major defeat in 1176.

Alexander returned to Rome, where in 1179 he convened the Third Lateran Council. The council issued the first major decree on papal elections since 1059. The earlier decree, assuming that elections would be unanimous, had made no provision for when a divided College could not agree on a candidate by consensus. The new decree eliminated the previous one's distinctions between the ranks of cardinals and required a two-thirds majority of all voting cardinals for election, if unanimity was impossible. Factions in disputed elections since 1059 had invoked the principle of the "better and sounder part." Now claims to be the "better part" were replaced by a numerical counting of the electors, becoming the first institutional example in European history of identifying a numerical majority. Excommunication was mandated for those who refused to accept a pope elected by two-thirds. The decree did provide for an election by less than two-thirds, but only if the Church harmoniously received a pope so elected. A pope elected by two-thirds of the cardinals received full power of the papacy upon accepting the office, except the right to ordain if he were not yet a bishop. From this point on papal elections have changed only in details. The two-thirds requirement meant that no faction in the conclave could elect a pope by itself, but it also meant that such a faction could prevent the election of someone it opposed. It would take three centuries before clear-cut parties emerged in the conclaves, but once they did, their role in the papal elections was mostly negative, since they usually prevented a strong candidate from becoming pope.

Two years after the council, Alexander III's long reign ended, and the first election under the new rules took place. The cardinals chose Lucius III unanimously, so the two-thirds rule was not invoked. Lucius died at Verona in 1185, and Urban III, the next pope, described the election there as being done with so harmonious a spirit that only God could have inspired it. Only eleven cardinals who had accompanied Lucius to Verona were present for the election. Urban, the archbishop of Milan, was so hostile to Barbarossa that he refused to give up that office after becoming pope in order to prevent the emperor from filling it. Urban's two-year reign was filled with conflict with the emperor, and he died in Ferrara fleeing the Imperial army. Three groups of cardinals each recommended a candidate, but one of those nominated said: "I tell you that I shall never accept this office, and the cardinal of Palestrina is entirely unequal to such a burden. It only remains, therefore, to elect the Cardinal of St. Lorenzo; for no one among us is so suitable, no one knows the customs and rights of the Roman church as he does."[8] Thus Gregory VIII was unanimously elected. The cardinals turned to the eighty-seven-year-old man, the oldest pope ever at election, because he was seen as conciliatory toward Barbarossa. He had only a two-month reign, but one of his acts shows the lack of charity that too often pervaded the medieval papacy. During a stopover in Lucca, he ordered the bones of antipope Victor IV thrown out of the church where he had been buried.

Gregory died a week later at Pisa, and the third successive election away from Rome took place, selecting Clement III, yet another unanimous choice but made by only eight cardinals. His death in 1191 occurred close enough to Rome that he could be buried in the Lateran, and the next election took place there. All thirty-one cardinals then alive chose their senior member, who became Celestine III. He was eighty-five at his election, but managed to reign for seven years, one of three popes who have lived past ninety. He maintained an active agenda until to the end, but late in 1197 he indicated his intention to resign if the cardinals agreed to elect his choice for successor. The cardinals objected to both his designating his successor and abdicating: "It was a thing unheard of for a pope to abdicate."[9] He died early the next year.

After the Mass of the Holy Spirit, the first time it is mentioned as being said before an election, twenty-one cardinals assembled in Septasolis. After a rancorous first ballot, in which those opposed to Celestine's designee emphasized the College's right to elect freely, a white dove, so the story goes, landed by the right hand of Lotario di Segni, a cardinal-deacon. He was elected unanimously on the second ballot. As was still the practice until the next century, the senior cardinal-deacon gave him his name, Innocent III. The first real test of the new papal electoral system put the perhaps most effective pope ever on St. Peter's throne. He was from a Roman noble family that had already had one pope, his uncle Clement III, and would provide three more. For all of his impact in history, however, he had no influence on the system of papal election. It had worked well for twenty years and six elections, and there was no need for change. Innocent died at the relatively young age of fifty-six at Perugia, with only two cardinals in his company. The rest of the College delegated them to choose Pope Honorius III, one of those in Perugia. Despite being over seventy years of age, he reigned for eleven years before dying in 1227 at Rome. The College, meeting again at

Septasolis and numbering eighteen cardinals, delegated three members to identify candidates. Their first choice refused the office, so they chose Innocent's nephew, who became Gregory IX.

Gregory made a competent pope, but he found himself entangled in the latest episode of the long-running Imperial-papal conflict. Emperor Frederick II had gained his throne through the goodwill of Innocent III, but his gratitude toward the papacy did not survive Innocent's death. By 1240 war had erupted in Italy between papal and Imperial forces, and the Imperial army blockaded Rome. When the pope called for a general council to condemn Frederick, the emperor captured the bishops and cardinals traveling to Rome. When Gregory died in 1241 in the city, there were only twelve cardinals, and two were captives. As happened in the past when Imperial-papal conflict raged, the college was divided between pro- and anti-Imperialists. A two-thirds majority could not be found for a valid election. The process dragged on through the heat of the late Roman summer. Ronaldo Orsini, head of Rome's government, locked the cardinals in a dilapidated building in hopes of forcing them into a decision, yet when it appeared that they would elect a noncardinal, he threatened them with violence. The building had inadequate facilities for the cardinals, as few as they were, and their retainers and servants. The latrines were not cleaned, and physicians were not permitted in to treat cardinals who fell ill. Finally, following the death of a cardinal, who allegedly was tossed into a casket before he was dead, and a threat from Romans that they would exhume the dead pope's corpse and have it make decisions, the cardinals agreed on the man who had the most votes on the first day. He took the name Celestine IV. Although the procedure of voting in a locked room did not become standard for papal elections for three more decades, it was the first conclave, since the word comes from the phrase locked *cum clave*, "with a key."

Unfortunately for the cardinals, Celestine was already terminally ill when elected and died seventeen days afterward at Anagni. His is one of the three shortest reigns of those officially recognized as popes. The cardinals had to go through the electoral process again, further handicapped by the loss of one member of each faction, so they were as deadlocked as before. The emperor could have ended the stalemate by releasing the two cardinals he held captive, but expecting that their votes would go to a candidate hostile to him, he refused. Only after eighteen months, with Frederick systemically pillaging the cardinals' properties, was a conclusion reached. Innocent IV, who took the name in memory of his powerful predecessor, made no concessions to Frederick II. He denounced the emperor as the Book of Revelation's beast rising from the sea, that is, the Antichrist. The emperor gave as good as he got; one of his entourage found that Innocent IV's name when decoded formed the number 666, the mark of the Antichrist.

Innocent's eleven-year reign provided him time to name fifteen cardinals; yet there were only eighteen upon his death in 1254 at Naples. There is no good explanation why most medieval popes failed to fill all the vacancies in the College and why some made no appointments at all even when there were many seats open. The factors involved may have included the expense of maintaining many cardinals and the diminishing prestige and revenues of each cardinal as their number increased. By 1300 the official size of the College was set at twenty-four, but for the next two centuries, it was usually in the teens.

The election of 1254 ended quickly compared to the two previous ones, only five days after Innocent IV's funeral, with a great-nephew of Innocent III chosen as Alexander IV. The conflict with the emperor continued unabated during his reign, but of greater significance for the next papal election was the fact that he appointed no cardinals during his seven years as pope. Their number had dropped to eight when they met in Viterbo in 1261. Hoping to win the long conflict against the Hohenstuffen heirs of Frederick II, who had died in 1250, with French aid, the cardinals elected a French bishop who happened to be at the papal court at the time. Taking the name Urban IV, he quickly named fourteen cardinals, including six from France. This was the largest non-Italian contingent of cardinals to date. The new French influence in the college manifested itself after Urban died in 1264 at Perugia, although it took four months of deadlock before the College turned to a French cardinal, who was a widower with two daughters. He was serving as legate in England and did not arrive in Perugia to be crowned until a month later. He became Clement IV. He, like Urban IV, never set foot in Rome as pope because of the hostile commune. His death in Viterbo in 1268 set the stage for what still is the most extraordinary papal election in history.

In opposing the Hohenstauffens, Clement IV had made a pact with the devil, in this case the French royal family. In 1265 he granted Charles of Anjou, brother of Louis IX, the throne of the kingdom of Naples, which was a papal fief. Charles invaded Naples with papal blessing and killed the last Hohenstauffen in 1268. He had greater ambitions than simply being king of Naples; he wanted to create an empire in Italy and beyond, for which he needed an allied papacy. When Clement died later in 1268, Charles acted to control the outcome of the election. When the cardinals met in Viterbo on December 1, 1268, the second day after Clement's funeral, there were twenty cardinals, but one was in France and never participated. Of the nineteen present, eleven have been identified as favoring the emperor, and the others, Charles of Anjou. It is clear from the sources that the cardinals came and went between the cathedral and their homes in the first months of the election process. After two months of no success, the cardinals were on the verge of electing Philip Benizi, the general of the Servite Order, who had gone to Viterbo to admonish them on the kind of pope the Church needed. He fled the city immediately, preventing his election. By late 1269, with no conclusion to the endless balloting (we do not know how often balloting occurred), the cardinals allowed themselves to be locked in a palace that had served as the papal residence for Clement. It failed to persuade them to agree on a pope, nor did strong letters from kings and princes across Europe.

As the conclave slipped into 1270, the Viterbese began to put greater pressure on the cardinals. The ultimate act in that regard occurred when the roof was taken off the palace in June. The Viterbese supposedly acted on the humorous comment of the English cardinal present that the roof should be removed to give free access to the Holy Spirit. This was the first time that the Spirit was noted as guiding the choice of a new pope. The cardinals threatened to put the whole city under interdict. A makeshift roof was put back on. The hardship of the long conclave resulted in two cardinals dying and a third leaving because of ill health. Charles of Anjou was accused of being behind the coercive tactics of the Viterbese. Whether or not he was, when he and Philip III of France visited the city in March 1271, it placed greater pressure on the cardinals to elect a pope. Finally, on September 1, they agreed to form a committee of six cardinals, three from each faction, with the power to elect, with five votes needed for the election to be valid. Bonaventure, superior of the Franciscans, admonishing the cardinals to finish, suggested Teobaldo Visconti, who was serving as legate to the crusaders in Syria. His election was announced on September 1, but he did not return to Italy for six more months. Gregory X was ordained a bishop and crowned at Rome on March 19, 1272. Forty months after Clement IV died, the Church finally had a hand at the rudder of the Bark of St. Peter.

CHAPTER III

The Creation of the Conclave

The incredibly long and difficult election of 1268–71 reflected the vast political power of the papacy. Neither party in the College of Cardinals was willing to concede to the other the authority and resources that having a pope on their side would bring, despite the hardship and scandal that their intransigence created. It was, however, the last gasp of the party that favored the Holy Roman emperor, made up mostly of older cardinals; the emperor no longer would be much of a factor in conclave politics. To be sure, later emperors sought to interfere in papal elections, but the threat to the independence of the College for most of the next two centuries came from the French monarchy. The French kings would prove more successful in shaping the papacy to their goals than the most powerful German emperors had ever been.

After his election Gregory X summoned a general council, which met at Lyon in 1274. A major issue was the method of electing the pope. The council approved of Gregory's proposals presented in his bull *Ubi Periculum*. It required that the conclave begin ten days after the pope's death in the same city where he died. The practice of displaying the dead pope's corpse for several days was recent; earlier popes had been buried on the day after they died, which was still true for Celestine IV in 1241. The bull required a novena (nine days of prayers) before the burial. If the city where the pope had died was under interdict or in rebellion, the conclave would take place in a nearby city.

The chief officers of the city where the conclave took place had the obligation to see that it was conducted properly. The cardinals would be locked into conclave with just one servant each, although two were permitted for special situations. Gregory mandated that all sleep together in one large room with only curtains to separate their beds. Once locked in they could not communicate in any fashion with the outside until the election was over, except with the agreement of all. A turnstile was set in a window to pass food inside. Only one plate of food and a bowl of soup would be provided per person, and after five days this would be reduced to bread, water, and a little wine. There was a timetable for reducing the cardinals' comfort level as the conclave dragged on. They could not conduct any business in conclave except for the defense of the Church. The papal *camerlengo,* who was rarely a cardinal before 1500, oversaw routine affairs during the conclave. Entry of tardy cardinals and reentry of those who exited for ill health were permitted. Imposing more urgency on the cardinals was the rule that they could not draw their incomes during a conclave. The conclave has changed only in details since 1274, although like previous election decrees, this one was suspended several times before becoming permanent. Gregory also conferred the new office of conclave marshal on the Savelli family as a reward for their valuable service in safeguarding the cardinals during the previous conclave. They held the office until the family's extinction in 1712.

Gregory died at Arezzo in 1276. The conclave that followed with ten cardinals was a model one. It required only one ballot, done in an hour's time, for the unanimous election of a saintly Dominican friar who was also cardinal-bishop of Ostia. His given name was Peter, but he took the name Innocent V. The promise of his reign was cut short by his sudden death at Rome only five months after his election. Again the cardinals assembled in conclave, but this time it took two weeks for the pope to be chosen. Charles of Anjou as Senator of Rome supervised the conclave and enforced the rules rigidly. Every door and window of the palace where they met was walled up except for the highest windows, through which, it was said, only a bird could gain entrance. After eight days he enforced the rule that restricted the cardinals to bread and water, although he allegedly slipped better provisions to the French. When he got word to the French contingent that Cardinal Ottobono, the choice of most

Italian cardinals, was acceptable to him, Ottobono was unanimously elected, becoming Hadrian V. His frail health was weakened further by the conclave. When his relatives congratulated him on becoming pope, he replied: "Why are you glad? A live cardinal can do more for you than a dead pope."[1] He died at Viterbo even before he was enthroned, but his election has been sufficient to have him included on the official list of popes.

Hadrian told the cardinals after his election that he would suspend the conclave rules and devise new ones, but his death occurred before anything official was done. The nine cardinals present, out of twelve, sought to avoid being subjected to them when they assembled for the election by swearing that Hadrian had suspended them, but the people of Viterbo would hear nothing of it and rigorously enforced the rules. Pedro Julião's election took place on the tenth day after Hadrian died, probably on the first ballot. The fourth pope of 1276 was both the only Portuguese and the only physician to be pope. He took the name John XXI, although there is no John XX on the official list. There was confusion about how many Popes John there had been in the tenth century. His first acts involved suspending Gregory X's rules of conclave and censuring those who had been overly zealous in applying them at Viterbo. John's eight-month reign came to a sudden end when the ceiling of his study in the palace at Viterbo fell on him.

For the fourth time in sixteen months the cardinals, now numbering only seven, found themselves confined in conclave. This time the process went far less smoothly. Charles of Anjou, who had interfered little in the previous three elections, reasserted himself. The conclave was deadlocked for six months between his allies and foes. For the first time the cardinals had to deal with several matters while in conclave that could not wait for it to end, for example, putting the city of Ascoli under interdict for attacking a papal castle. The deadlock was finally broken when a member of the Orsini of Rome was chosen as Nicholas III. The Orsini were then the most powerful Roman family and would remain a potent force in Roman and papal politics for five centuries. They had already had one pope in the family, Celestine III. Nicholas was an active pope, rebuilding the Vatican palace to have a residence close to St. Peter's and working to arrange a peace between Charles of Anjou and Rudolf, the first Habsburg emperor. His close ties with Rudolf secured imperial recognition of papal rule of the Romagna, which included Bologna. This set the borders of the Papal States largely as they were until 1848. Nicholas's reign was quite brief. He died suddenly in 1280 at Viterbo, which was the favorite residence of popes in the thirteenth century and the site of more elections than Rome.

The day after his election, Nicholas III had suspended the rules of conclave from 1274, but it made little difference as the election of 1280–81 also dragged on for six months. The thirteen cardinals, who met in the palace of Santa Sabina, were sharply divided between five who favored the Orsini and eight in Charles of Anjou's party. The latter regarded Nicholas's policy as too favorable to the emperor. The Angevins were one vote short of two-thirds, which seemed out of reach until the Viterbese acted after five months. Nicholas had appointed as the mayor of Viterbo a relative who had badly alienated its people. When in early February 1281 rumors spread that an Orsini would be elected, they attacked the palace and dragged out the two Orsini cardinals. One was soon released, but the other was imprisoned for the duration of the election. His abduction gave the pro-French party a two-thirds majority, and Simon de Brion, a French cardinal, was elected. He took the name Martin after Saint Martin of Tours. At the time of his election it was believed that Marinus I and II were Martins, and so he became Martin IV, although he was in fact the second Pope Martin. One can assume that he was pleased that the violence of the Viterbese had given him the papacy but angered at their affront to the dignity of the cardinals. He moved his coronation to Orvieto, because Rome refused the new pope entry. A month after his election, rebellion broke out in Sicily against Charles of Anjou, which drove the French out. The Sicilians offered the rule of the island to Martin, who refused it and ordered them to return to French rule under pain of excommunication. The Sicilians then turned to King Peter of Aragon, who accepted, defying excommunication. Thus, Aragon also became involved in Italy.

Much of Martin's work as pope involved aiding French interests. He canonized Louis IX and appointed mostly French cardinals. There was a hostile reaction from other kingdoms and Italian cardinals. When Martin died in March 1285 at Perugia, there were eighteen cardinals, but three were away serving as legates. All fifteen cardinals present elected an elderly Italian cardinal who was the grandnephew of Honorius III. To honor him he became Honorius IV. As frequently has happened in the history of papal elections, the new pope was intended as a caretaker, who would hold the office for a short time, while the factions took the time to strengthen themselves for the next election. Honorius, unlike many other caretaker popes who proved vigorous and active, met expectations by doing little and dying in two years. He was, however, a native of Rome whom the Romans welcomed back to the city. He was the first pope in decades to spend his reign in Rome, where he died in April 1287.

The next election took place again in Santa Sabina. It was as bitter and drawn-out as the previous one had been brief and harmonious. Compounding the divisions among the fifteen cardinals was that fact that nearly every cardinal had ambitions to become pope, which has been a problem in papal elections less often than one might expect. Although Charles of Anjou had died two years earlier, his son Charles II of Naples interfered as much as he had. Without the harsh rules of conclave to push them to make a quick decision, the cardinals haggled into the hottest summer in memory at Rome. Six cardinals died before the election was suspended until the winter. Even after the cardinals returned to Rome, it took until February 1288 for the pieces suddenly to fall into place. The nine surviving cardinals unanimously chose one of their own, the general of the Franciscans, the only cardinal who had remained in Rome. He at first refused on the grounds that a Friar Minor had no business being pope, but when a week later he was again elected, he accepted. He took the name Nicholas IV. His reign is noteworthy for sending Franciscan missionaries to the Mongols and the loss of Acre, the last crusader stronghold in Palestine. Nicholas died in 1292 at Rome.

Ten cardinals assembled at Rome in April 1292, to begin deliberations. Nicholas had appointed only one cardinal, Benedetto Caetani, so the College was smaller than in 1287. There were two Orsini cardinals, whose family had changed its policy to favor the French in southern Italy. Another Roman family, the Colonna, bitter rivals to the Orsini, also had two cardinals, who championed the Aragonese in Sicily. King James of Aragon provided gold for bribing the cardinals, but it is not known whether it was actually distributed. After ten days of balloting, no candidate was close to the required two-thirds of the votes. Hoping that a change of scenery might help, the cardinals changed their residence and tried again in June. An epidemic broke out in the city during the summer, and a cardinal died in August. The cardinals dispersed to more healthy locales, and they reunited in late fall with no greater success. The balloting continued into the next summer. Since the cardinals were not under the rules of conclave, they carried on their routine while the papacy was vacant. How many ballots were taken is unknown.

Meanwhile the city of Rome was rife with violence and disorder. Until 1700 all prisoners were released at a pope's death in imitation of Barabas's release at the time of the crucifixion, and many returned immediately to the life of crime that had landed them in prison. During the *sede vacante,* the *camerlengo* would issue an *adviso* that laid out precisely the rules of behavior and draconian punishments for violations, which were placed on placards and posted around the city. Rarely did it have much effect. The papal interregnums were usually marked by violence up to 1700, but few surpassed the disorder of 1292.

After dispersing again early in the summer of 1293, the cardinals agreed to go to Perugia in the fall. This failed to secure an outcome for the balloting that continued through the winter. In March 1294 King Charles II of Naples arrived in Perugia. Seated between the two senior cardinals, he addressed the cardinals on the need to elect a pope without delay. Caetani, suspecting that Charles was pressing the College to elect his own partisan, made a sharp rejoinder on how the election of a pope must be free and without coercion. The king went on his way without seeing any results.

As the process dragged into its third summer, several cardinals left for the summer, leaving only six in Perugia. At a final meeting before disbanding, one of them read a letter from a hermit, Pietro de Murronne, declaring that God had revealed to him that the cardinals would be severely punished if they did not elect a pope immediately. He was well known to the cardinals as a saintly miracle worker, and the senior cardinal suddenly nominated him as pope. The others quickly agreed, and those who had left were recalled to give their consent. On July 5, 1294, two years and three months after

Nicholas IV had died, all eleven cardinals elected Pietro. His election was deemed done by "inspiration."

Pietro, who was possibly as old as eighty-six, was truly shocked upon being informed of his election as pope, but, concluding that it was God's will, he agreed to be consecrated as Celestine V. He rode a donkey to the nearest cathedral for enthronement. His hermitage was near Naples, and, encouraged by Charles II, he took up residence in that city. He filled the College by appointing thirteen cardinals in short order. Seven of them were French; two Italians were partisans of Charles II; and three came from his own order. Celestine found himself overwhelmed by the burdens of his office, for which he was utterly unprepared in every respect. He began to think of abdicating. Caetani, who was his principal adviser, was later accused of coercing him to abdicate. Celestine's last act reimposed the rules of conclave of 1274, which have remained in effect, albeit with some changes, to the present. Hearing that the cardinals hoped to evade a strict conclave on the grounds that it applied only at a pope's death, he further decreed that the rules of 1274 were in effect however a vacancy might occur in the papacy. Then on December 13, 1294, before the entire College, he read the act of abdication that Caetani had prepared for him and urged the cardinals to proceed immediately to conclave. They unanimously agreed to accept his abdication. Pietro expected to return to his hermitage, but his successor, fearful that he would be made the center of opposition to him, had him confined to a castle tower. He died there in May 1296; whether he was murdered to remove a threat to the new pope remains uncertain.

Few events in papal history have been as controversial as Celestine's election and abdication. Many, especially the radical Franciscans known as the Spirituals, had viewed his election as the coming of the Angelic Pope who would reform the Church before Christ's Second Coming. The Angelic Pope was an idea that had first appeared in the Imperial party during the bitter feud between emperor and pope in the twelfth century. It had spread among those who felt that the Church and the clergy were too worldly and hoped that a pope would appear to purge the Church of its property and concerns with wealth and political power. Celestine's resignation deeply embittered them, but, rather than blaming him, they accused Caetani of coercing him.

Caetani became Pope Boniface VIII in the brief conclave at Naples that followed the abdication of Celestine, who reportedly urged the cardinals to vote for him. For the first time there exists a document that describes the conclave in some detail as it actually happened.[2] It began the required ten days later and lasted only one day. Charles II of Naples shut the voters up in cells only ten feet square and allowed them only one servant; previous conclaves had been less austere. The cardinals wrote on their ballots: "I, . . . , Bishop of . . . , elect as Supreme Pontiff Venerable Father my Lord the Cardinal of. . . ." Three cardinals were chosen as scrutators to count the ballots, and three more served to scrutinize the scrutators to ensure their accuracy. While there was no rule against voting for one's self, a two-thirds majority plus one was required to ensure that a cardinal's vote for himself did not provide the margin of victory, a rule not always followed in the future. While the ballots were counted, no comments on the merits of those receiving votes were permitted. The cardinal chosen on the first scrutiny, as the process of balloting was called, refused the office; a second one was inconclusive; Caetani was unanimously elected on the third. The new pope, a renowned canon lawyer, nullified Celestine's last acts, except for the one reimposing strict conclave. He went to Rome, where he was crowned a month later.

Boniface VIII enjoyed the power, prestige, and wealth attached to the papacy. In that respect he was the opposite of the Angelic Pope Celestine, and those who regarded Celestine as a saint were hostile to Boniface from the first. By favoring his relatives he angered the Colonna, whose two cardinals turned implacably against him. Boniface banned them from the next conclave. Later popes would react against this blatantly political act by declaring that nothing except apostasy or heresy could prevent a duly appointed and mentally competent cardinal from voting for pope or receiving votes.

Almost immediately, protests arose that Boniface's election was invalid because a pope was not permitted to step down. What should have been only a footnote in papal history became a major controversy because of Boniface's conflict with the French monarchy. He began his reign by supporting the French cause in Sicily, but Philip IV, cunning politician that he was, sensed an opportunity to extract greater favors from the papacy and asked the University of Paris to issue a judgment on whether a pope could abdicate. The Paris theologians decided that he could not. Beyond the lack of any examples of a pope abdicating except under severe coercion, they placed a great deal of weight on the claim that a bishop had a spiritual marriage with his diocese. Like a carnal marriage, it could not be broken. The Council of Nicaea had used this analogy to justify prohibiting bishops from moving from one see to another. Although the rule had been ignored for centuries, it was used to proclaim that Celestine could not rightfully abdicate and that Boniface's election was invalid.

Having gained the decision he wanted from the university, Philip intended to leave it unused until he might need it, but the moment arrived sooner than anyone could have expected. France and England were at war over Gascony, which Eleanor of Aquitaine had passed to her second husband, the Anglo-Norman ruler Henry II, and both sides were funding the war by taxing the clergy. The Fourth Lateran Council had decreed as law the principle that the secular rulers could not tax the clergy without papal consent. In 1296 Boniface, who was adamant about papal authority and clerical autonomy, denounced both kings for their actions. When in 1301 Philip imprisoned a bishop for treason, Boniface issued his *Unam Sanctam,* notorious for its claim of papal authority over rulers. Meanwhile the Colonna cardinals had been driven out of Rome, and they and other foes of the pope arrived at the French court eager to persuade Philip to act against him. The French estates-general met to support the king against the pope. It denounced Boniface for heresy, simony, and Celestine V's murder and declared that he was an antipope, because Celestine could not validly abdicate. Boniface intended to excommunicate Philip; but before he did, French troops and the Colonna seized him at Anagni, his hometown, in September 1303. The locals soon freed him, but the rough treatment and humiliation inflicted on him broke his spirit, and he died a month later at Rome, where the Orsini had taken him. The day after Boniface was buried, Charles of Naples, who controlled Rome with his army, locked the cardinals in conclave at the Lateran.

In 1303 the events of Boniface's last days forced the cardinals to think only of electing someone who would not anger Philip. On the first ballot a cardinal who was a Dominican was unanimously elected. By becoming Benedict XI, he indicated his belief that Benedict X, deposed in 1056, was a true pope and not an antipope as he is listed in the official list of popes. While an Italian, Benedict XI was not a Roman, so he was not encumbered by deadly feuds between powerful Roman families; nor was he marked by the struggle between Philip and Boniface, since he had been largely absent from the curia. Benedict came from a humble family, but unlike Celestine he was an experienced churchman. His brief reign saw accusations that he was too favorable to his order, which provided all three of his cardinals. Backtracking from Boniface's positions, he refused to publish the bull excommunicating Philip and the Colonna, although he did not restore to the Colonna their properties stripped by Boniface, and he retained the excommunication of the French leader of the attack at Anagni.

The Romans had supported Boniface, and Benedict's acts offended them. They drove him out of the city. He went to Perugia, dying there in July 1304. Nineteen cardinals assembled at Perugia for the next conclave. Benedict had been the only neutral cardinal the year before, leaving none for this election. There were fifteen Italian cardinals, but Orsini and four other Italians had signed on to the French cause. They and the three Frenchmen prevented the election of any Italian who might seek revenge for Boniface. The conclave lasted until June 1305, and four cardinals were forced to leave because of illness. Well before June it was clear that the cardinals had to go outside the College to find a pope, and after several false starts, they found him in Bertrand de Got, archbishop of Bordeaux. He was French, yet his diocese was in Gascony, a region subject to the English king. He had traveled to Rome in 1302 when summoned by Boniface for a synod, although the king had barred the French from attending. Thus he was perceived as neutral, not the lackey of Philip IV that he soon became. A cardinal of the pro-Boniface faction had first suggested Got's name, and the French quickly rejected him. Orsini had second thoughts and asked Philip's opinion. The king decided that he could support the election of Got, who received ten of fifteen votes. He took the name Clement V to honor the French pope Clement IV.

Because Clement was in France when elected, he was crowned five months later in Lyon, so Philip could be present. Within a month he had reinstated the two Colonna cardinals and appointed ten cardinals, nine of them French, including four nephews; the tenth was English. Thus began a pattern that marked the papacy for the next seventy years. Of the 136 cardinals created in that era, at least 112 were natives of the lands within the kingdom of France, although not all were the French king's partisans. Clement took his time about going to Rome, but in 1309 he began the trek to Italy. Falling ill he stopped at Avignon to recover. Informed that violence in Rome was making it unlivable, he decided to remain in Avignon. The city was French in culture but not a part of Philip IV's realm, since it was in Provence, which was ruled by Charles of Naples. As a papal vassal Charles was obliged to be amenable to the papal will. Avignon also bordered on the Venaissin, a small county that had been given to the papacy in 1229. Its location on the lower Rhône made Avignon accessible from northern Europe as well as the Mediterranean. It was closer to the center of Catholic Europe than was Rome. Clement V established himself in Avignon's cathedral, although he never openly abandoned plans to go to Rome.

The five years that Clement spent at Avignon demonstrated the clout the French king could exercise over the papacy when it was that close to him. Certainly the new pope would have favored the French king in any circumstances, but if he had been at Rome, French influence surely would have been less. Philip prevailed on Clement to lift the excommunication of the leader of the attack on Boniface VIII, praise Philip for his opposition to Boniface, and canonize Celestine V. More notoriously Philip also persuaded Clement to disband the Knights Templar, whose wealth Philip intended to seize, and condemn its leaders for heresy and multiple other crimes and sins. Legend has it that when Jacques de Molay, the Templar grandmaster, was burned at the stake, he called down a curse with his dying breath that within a year both pope and king would be before God's judgment seat. First pope and then king died in 1314.

Clement V died at Carpentras in the Venaissin, and the election that took place in the bishop's palace there was the first one outside of Italy since the system of conclave had been created. The twenty-three cardinals were mostly French, but they were divided between clients of the French monarch on one side and, on the other, Clement's relatives and other Gascons, who had to answer to the English king as their feudal lord. Nor did the seven Italians form a unified bloc. As the conclave dragged into the summer, troops allegedly in the pay of the dead pope's relatives broke into the conclave shouting, "Death to the Italians!" The Italian cardinals fled to Avignon; the rest joined them several months later. They still failed to agree on a pope. French King Louis X persuaded them to move to Lyon. In August 1316, eighteen cardinals finally agreed to elect a former archbishop of Avignon as John XXII. He happily settled back in his old haunts. He had promised the Italian cardinals to return to Rome, but he made no effort to do it beyond sending an army to reassert his control over the Papal States. John became notorious for condemning Franciscan apostolic poverty as heresy.

John believed that papal authority included the right to appoint the Holy Roman emperor. In 1314 there had been a double election, and both candidates appealed to the papacy. John decided against Louis of Bavaria, who proceeded to defeat his rival and become the center of resistance to the pope's policies. In 1328 Louis went to Rome, where he was crowned emperor and named a Spiritual Franciscan as pope, who called himself Nicholas V. Louis and his pope held a synod in which John was convicted of heresy and deposed and burned in effigy. Nicholas soon repented and three years later went to Avignon to seek forgiveness. John, who usually was harsh to his foes, forgave him and exacted no further penalty from him. Nicholas was the first antipope in 150 years. The long period without one was largely a result of the reduction in power of the Holy Roman emperor, but it also reflected the success of the papal electoral system in place since 1179.

John XXII was about seventy-two years of age when elected. Like many old popes, he fooled his electors and had a long reign. He reigned for eighteen years and probably was ninety at his death in December 1334. John was a shameless nepotist and filled the College with his relatives, resulting in more cardinals than there had been for over a century. Twenty-four cardinals went into conclave in Avignon. The dominant French sought an oath from their first candidate that he would never return to Rome, but he refused to give it. They turned to Jacques Fournier, bishop of Pamiers, who garnered some Italian support and was elected, although it appears that he received sixteen votes, exactly two-thirds, and not the two-thirds plus one that was expected for election. It is not known whether Fourier, who became Benedict XII, actually pledged to remain in Avignon, but he began building the great papal palace there, at least suggesting he had no intention of returning to Rome. From a humble background, Benedict had been a successful inquisitor in southern France.[3] As pope, he turned his inquisitorial zeal and sense of duty toward rooting out corruption in the Church, especially in the papal court and the monastic orders. Unlike his two immediate predecessors, he refused to appoint his relatives as cardinals, saying that a pope must be without kin. His attacks on corruption earned him bitter enemies among the cardinals, even those he appointed—five Frenchmen and one Italian.

When Benedict died at Avignon in 1342, he left no party in the College to perpetuate his policies. Nineteen cardinals were close enough to Avignon to participate when the conclave began ten days after his funeral. Several cardinals were elsewhere and did not take part. Few papal elections from this point on had all cardinals present. The election of 1342 provides evidence of the pattern where cardinals would choose a man whose personality and policies were far different from those of the dead pope. Nothing is known of the activity within the conclave.[4] Cardinal Pierre Roger, the archbishop of Rouen, was elected unanimously after four days. The cardinals wanted him because he was quite the opposite of Benedict—affable, luxury-loving, and tolerant of human foibles—although he also was a good theologian. French King Philip VI sent a letter to the cardinals urging his election, but he had already been chosen when the letter arrived. Philip wanted him as pope because he was a product of the royal court, serving as royal chancellor before becoming an archbishop. It was becoming common for clerics to rise in the hierarchy through royal service rather than in the Church. Roger took the name Clement VI.

A major reason the French monarchy pushed the cardinals on the papal election was the war with England. The Hundred Years War had broken out five years earlier. Benedict had been as neutral as possible for someone who had French blood in him, but Clement VI openly supported the French cause. His devotion to the French monarchy was the principal reason the Avignon popes gained their reputation as being subservient to French interests. Along with extravagant spending on the papal court, Clement made large loans to the French king, which required raising greater revenues for the papacy. Papal financial exactions led to bitter resentment in England, where Edward III angrily snapped that Jesus had told Peter to feed his sheep, not shear them. In 1351 Edward issued the first of a series of statutes that reduced papal rights in England and the flow of English money to the papacy. They would eventually be ignored, as England returned to union with the papacy after 1378, but they remained on the books for Henry VIII to use 150 years later.

Six months after Clement became pope, a delegation came from Rome to beg him to return. He responded by purchasing Avignon in 1348 for a small sum from Queen Joanna I of Naples, who was also countess of Provence, showing that he expected residency there to be permanent. To his credit, when the Black Death hit Avignon the same year, he defended the local Jews against charges that they had caused it by poisoning the wells. It probably was not the plague that caused Clement's death in 1352, since the disease had died out in Avignon by late 1349, having killed six cardinals. Clement had actively erected cardinals, who numbered twenty-eight in 1350, the most in two centuries. Twenty-five cardinals assembled in the papal palace for the next conclave. The cardinals had prevailed on Clement—they did not have to push him hard on it—to reduce the severity of conclave. They were allowed to have two servants instead of one, and their meals were greatly improved, with more and a wider range of foods and delicacies permitted, including salads, sweets, cheeses, and fruit.

Ironically the first conclave in which this relaxation took effect was one of the shortest in a century, demonstrating that the severity of the conclave rules was far from the most important factor in determining how quickly the cardinals would elect a pope. The cardinals first drew up a document called a capitulation intended to enhance their power vis-à-vis the pope. The number of cardinals was to be limited to twenty, and the next pope could not create new cardinals until the number had dropped to sixteen. A two-thirds majority of the College had to approve of creating cardinals, excommunicating them, or depriving them of the right to vote in a conclave. The pope needed the same majority to reduce a cardinal's properties and revenues. Approval also was required to request any subsidies from kings and national clergies. Further clauses gave the College veto power over numerous papal decisions and policies. Finally the pope was obliged to share equally all revenues with the college. Every cardinal swore an oath pledging their support of the capitulation should he be elected, although some added a hedge invalidating any clause inconsistent with Scripture. This was the first time that the cardinals sought to use the conclave to impose controls on the pope. Most conclaves for the next 300 years produced a written capitulation, and it is assumed that there were oral agreements in those for which a written one does not exist. In either case their impact was negligible.

Once the capitulation was written, it took only two days to elect the new pope. Yet there was some drama in the conclave. There was at first a push to elect the general of the Carthusians, an austere order; it was derailed by a cardinal's impassioned speech against selecting another Celestine V. The cardinals turned to Etienne Aubert, who was regarded as an eager supporter of the capitulation. He became Pope Innocent VI, because he was known to be acceptable to the French king, who did not arrive in time for the conclave, but attended his coronation. Innocent disappointed him by not being as fervent a supporter of the French cause as Clement had been. In 1353 Innocent declared invalid the conclave capitulation he had sworn to accept, because the cardinals had violated Gregory X's edict barring any business from being transacted in conclave except the election. He compensated the cardinals with increased revenues. This set a pattern repeated in numerous conclaves: The cardinals would impose conditions on the new pope only to have him reject them once he was in power.

A major argument against returning to Rome was the violence that pervaded the city, but by 1357 Avignon no longer was so safe. The English had captured the French king in 1356, which led to a hiatus in the Hundred Years War, and thousands of troops were set loose in France. Some of them attacked Avignon in May 1357. They were not enough to enter the city, but the event frightened Innocent. He began to think seriously about returning to Rome and sent an army to regain control of the Papal States to prepare for his return. The expedition accomplished nothing except nearly bankrupting the papacy. The pope died at Avignon five years later.

When the conclave began on September 22, 1362, there were several factions among the twenty cardinals; the largest was from the region around the French city of Limoges, with twelve members, including six relatives of Clement VI. They were two votes shy of what was needed to elect. The Limousins and another faction independently of each other chose to vote for Clement VI's brother, who received fifteen votes on the first day. He refused the office, saying that his health would not permit him to serve. The balloting that followed saw votes scattered among many cardinals, and no one gained strength as it continued. By the sixth day it was clear that no cardinal was likely to receive the required majority. Men outside the College were discussed, and a Limousin cardinal put forward the name of a French Benedictine serving as papal legate in Naples. He was elected on September 28. Fearing that if the Italians got word of his election, they would detain him in Rome, the cardinals sent him a message to come to them for consultation. It took him five weeks to reach Avignon, where he was crowned as Urban V. He was an austere monk who continued to wear his monastic garb as pope and reduced the luxury and frivolity of the papal court. Urban's most noteworthy act involved his trip to Rome in 1367. Most cardinals raised loud objections to the trip, threatening to make him go by himself. Urban silenced them by naming a twenty-eight-year-old cleric standing nearby as a cardinal and saying that he would appoint many more in the same manner if they refused to go to Rome. He remained thirty months in Rome but did not become fond of it. He and the cardinals were pleased to return to Avignon in September 1370, where he died in December.

On the tenth day after Urban's funeral, nineteen cardinals entered conclave. The fifteen French cardinals agreed without hesitation on Clement VI's nephew, one of five of Clement's relatives in the College; the rest voted for him once it was clear that he had the required majority. He took the name Gregory XI. At age forty-two he was still a deacon and had to be ordained a priest and bishop before receiving consecration as pope. There is no sense during the conclave or soon after that he planned to return to Rome, but the chorus of complaints about the pope's absence from the Eternal City was growing louder. The fact that several popes had not set foot in Rome was not in itself scandalous; it had happened often in the past, although it was true that those popes continued to live in Italy. It has been calculated that in the 200 years before Clement V's election in 1304, the popes resided outside of Rome for 122 years or 61 percent of the time.[5] The scandal lay in that the Avignon popes gave no hint they would ever return to the city where they were bishops and from which they drew their authority over the entire Catholic Church. Two strong-willed women led the chorus demanding the return to Rome. Bridget of Sweden, a princess, and Catherine of Siena, daughter of a dyer, were both mystics whose visions revealed the damage to the Church of the pope's absence from Rome. Both demanded that the pope return to Rome and used the harshest language in telling him. Bridget, who was in Rome while Urban V was there, was heartbroken when he returned to Avignon and had accurately predicted his early death if he went back.

Unlike Bridget, who died in 1373, Catherine saw the vindication of her cause. In early 1376 she went to Avignon as part of a Florentine embassy and spent half a year working on Gregory's resolve to stay in Avignon. In September the papal party began the trip to Rome, despite the fierce opposition of most cardinals, and arrived in January. Gregory found Rome not the paradise that Catherine had portrayed. The city continued to be wracked with violence, and its buildings were in serious disrepair. Forced to spend most of the summer of 1377 there to deal with the great problems that needed attention, Gregory's health was undermined. He allegedly was planning to return to Avignon. Before dying in March 1378, he told the cardinals to begin the election quickly after he died and ignore the rules of conclave if necessary in order to prevent factional coercion. He gave them permission to meet outside of Rome and to move the conclave many times if necessary. He also suspended the two-thirds majority rule, but his phrase "the greater part" is vague in what it required. Unfortunately, the cardinals ignored what proved to be very sound advice.[6]

Because of the events marring the conclave of 1378, the sources for it are more numerous and detailed than for any previous one, but also contradictory. They make it possible to describe a conclave for the first time in detail, although this one was so irregular that few useful insights can be drawn from it. Sixteen cardinals were in Rome for the election; six stayed in Avignon. The electors were eleven Frenchmen, four Italians, and one Spaniard. The

French contingent included eight Limousins, who formed a bloc, but the other three Frenchmen were bitterly opposed to them. Delayed for a day by a violent storm, the cardinals entered conclave in the Vatican palace on April 7, 1378. It was the first conclave to take place there, because the Lateran was unusable. As they pushed their way in, a crowd estimated at 20,000, including peasants from outside Rome, shouted: "We want a Roman pope, or at least an Italian, or else you will die!" One voice was heard above the rest: "If you cardinals don't give us one, we will make your heads as red as your hats!" Several French cardinals wanted to flee Rome and meet elsewhere, but the majority persuaded them to stay, arguing that they would be in greater danger if they tried to leave. The cardinals' servants had a great deal of trouble getting the palace cleared of those not permitted to stay; it was past midnight before the doors could be locked.

Early the next morning, while the cardinals attended the Mass of the Holy Spirit that began the election, they heard bells ringing across the city. The Italians among them recognized the sound for what it was—a call to arms. A huge mob assembled outside the Vatican demanding a Roman pope. Cardinal Orsini spoke to the mob from a window, saying that a pope elected under threat of violence could not be accepted as pope, but he failed to convince it to disperse. The cardinals urgently discussed their options. The possibility of schism was on their minds, as one pointed out that if they caved in to the mob's demand, the cardinals still in Avignon would probably choose their own pope. Orsini suggested dressing up a friar as pope and proclaiming him as such to the mob while the cardinals slipped out of Rome and held the real conclave elsewhere. That was deemed too undignified. As other options were considered and rejected and the mob grew louder, it was agreed to elect, or make a pretense of electing, Bartolomeo Prignano, the Italian archbishop of Bari. He was well known to the cardinals as an efficient and humble bureaucrat, since he served as vice-chancellor of the curia for twenty years.

The cardinals remained fearful that the Roman mob would not settle for an Italian instead of a Roman, so they propped up in a window the elderly, sickly Roman cardinal who was the senior member and announced him as pope. The cardinals did not want to elect him for real, despite the likely brevity of his reign, because it would appear that they had given in too completely to the mob. Much of the mob dashed off to his house to plunder it. This is the first mention of the practice of sacking the house of the pope-elect, although the Romans had plundered the belongings of popes when they died in the city. That had still occurred in 1227 when Honorius III died in Rome, the last pope to die there until 1378. Now, the practice shifted from plundering the deceased pope's possessions to the new one's. Its justi-fication came from a ritual in the papal coronation, when the new pope tossed three handfuls of coins to the crowd with the words: "Silver and gold are nothing to me. What I have I give you." Often during a conclave false rumors about a cardinal being elected led to the sacking of his house. Not only would a cardinal in such a situation not become pope, he would also emerge from the conclave to find his property ransacked. Numerous prohi-bitions against the practice failed to stop it until the late eighteenth century.

The cardinals voted again for Prignano, and he took the name Urban VI. They made their obeisance to him and gave him gifts. They then ac-companied him to the Lateran basilica to take possession of it. Nine days later they sent a letter to the cardinals in Avignon declaring that they had elected him freely and unanimously. There are two alternate scenarios about Urban's election. One is that the cardinals expected that he could be per-suaded to renounce the election once the mob had been pacified by the news an Italian had been elected, allowing the cardinals to slip out of Rome and elect another. The second is that the cardinals were serious in choosing him but expected him to be pliable in their hands, believing that since he had lived for years in Avignon, he would be willing to return there. Neither proved to be true. Having been raised to power far beyond his dreams, he was not about to let it slip through his fingers. He also revealed a far dif-ferent personality from the docile bureaucrat that the cardinals knew. When Orsini told him that the College had not intended for him to be pope and expected him to refuse the office, he flew into a rage. The Ro-mans, despite finding out that they had been duped about getting a Roman pope, accepted Urban and supported his right to the papacy. A week later he was enthroned at St. Peter's. Soon the cardinals found themselves the tar-gets of his fits of rage and slipped out of Rome. Within two months, twelve cardinals had fled, leaving four in Rome.

The cardinals had to face the fact that they elected a pope who in their minds was quite mad. Whether Urban truly was insane, as many commentators on the matter have insisted, is impossible to say. He acted imperiously and impetuously and, in respect to those cardinals who had refuted him, with rage and malice. The cardinals declared that they had not intended to make him pope and that the coercion of the Roman mob made the election invalid. In two months three more cardinals—the fourth was too ill to travel—joined their colleagues. On August 2, 1378, they formally declared the papacy vacant and ordered Urban to stop using the title of pope. They proceeded to a new election at Fondi. The French cardinals told each of the three Italians that he would be their choice, and during the election the Italians abstained, each expecting to be the one chosen. They were double-crossed, as the French all voted for the Cardinal of Geneva, who could thus claim to be unanimously elected. He called himself Clement VII. Over the next months Urban and Clement excommunicated each other and the cardinals in their parties. Urban had created his own cardinals. Although he quickly antagonized most of them as well, he always retained the support of the Roman populace. When it became clear that Urban was not going to submit and allow Clement to reside in Rome, Clement went back to Avignon, arriving there in June 1379.

For the first time, the same group of cardinals had erected two men as pope. In previous schisms, there had always been a division in the College. There was legitimate reason to accept either Urban or Clement as the rightful pope, and when they died, their cardinals proceeded to elect successors. In 1378 and for four decades to come, no one in Catholic Europe could find a resolution to this incredible problem.

CHAPTER IV

Conclaves during The Great Schism

In 1378 the papal electoral system had failed in the worst possible way, providing two men with strong claims to be the rightful pope. It was a failure first of all of the Avignon popes, who had ignored their obligation to reside in Rome. It was a failure too of the Romans, whose need for the economic benefits the presence of the pope provided them trumped the obligation of holding an orderly and proper conclave. It was also a consequence of papal rule over Rome. Had a powerful secular ruler been in control of the city, it is improbable that he would have tolerated so disgraceful an event. Fourth, it was the failure of the electoral process. By requiring the cardinals to hold the election in the place where the previous pope had died, the rules of conclave created the circumstances in which it could be questioned whether Urban VI was rightfully elected. Last, it was a failure of canon law, which declared that the pope could not be judged by any earthly authority. Any attempts at mediation by rulers or even by the College was deemed unlawful unless both popes agreed to accept arbitration, which neither did.

In short order canon lawyers proposed three ways that the crisis could be resolved. The most obvious was resignation by both popes, called the "way of cession," in which the legitimate pope, whoever he was, would step down. A new election by the cardinals of both factions then would choose a new and valid pope. The "way of action" was a second option. It referred to the use of force or at least diplomatic pressure on behalf of one pope to persuade the opposing one to concede. Medieval churchmen had no reservations about using violence in matters of religion to force an erring person to repent: Error has no rights. One reason both popes invested so much energy in trying to secure the obedience of the Catholic rulers was in order to use their military forces against the other. Neither succeeded because their supporters were about equal in strength. Most of Italy, the Holy Roman Empire, and England and its allies in the Hundred Years War sided with Urban. France, Naples, Scotland, and the Iberian realms were with Clement VII. Portugal seems to have been entirely neutral. There was some movement from one camp to the other among some small states, but the key issue was the war between England and France, which made England unwavering in support of the Roman pope. The third option was the "way of the council." Most thoughtful churchmen saw a meeting of a general council as the only solution, but there was a major problem: Only a pope could convene one. Neither pope was willing to take the chance of being declared an antipope by a council.

With both popes intransigent, the two Colleges regarded themselves as the proper body for electing the next pope. When the two contending popes died, their cardinals tried to follow the rules of conclave precisely to prove their legitimacy. Urban VI was the first to die, after a fall from a mule as he fled from violence in Rome in 1389. Few of the sixty cardinals he had created since 1378 were still alive or at his side. Six had been arrested for allegedly plotting to depose him in 1385 and were subjected to gruesome tortures. The Englishman among them was released when his king threatened to withdraw from obedience to Urban, but the other five disappeared and surely were executed. Two more of his cardinals fled to Avignon. Fourteen were in Rome when he died. The opportunity of recognizing Clement at Avignon and resolving the schism was passed over. Urban's cardinals believed themselves the valid electoral body, nor were they eager to test the generosity of their rivals in respect to keeping their rank and privileges. To prevent any interference with their election, they proceeded into conclave the day after Urban's death, but that was the only known point on which they violated the rules of conclave. By the time the news of Urban's death had reached Avignon ten days later, they had unanimously elected a cardi-

nal from Naples, who took the name Boniface IX. He was amiable compared to Urban, but he refused to concede anything.

Five years later, Clement VII died at Avignon only hours after falling ill.
The twenty-one cardinals waited the required ten days to enter conclave. A
courier had rushed the news of Clement's death to the French king in six
days; it took only four to get a letter from him back to Avignon. The courier
arrived just as the last cardinal was entering the papal palace and gave the letter to him. The cardinals debated whether they should read it, sure that it
was an urgent request from Charles VI that they delay the election and negotiate with Rome. They voted not to, on the grounds that most cardinals
were in conclave when it arrived, and the rules forbade any outside communications. The cardinals also debated whether to delay a vote until contact could be made with Boniface at Rome. They decided against it on the
basis that their late master had excommunicated him and his cardinals, and
they had to elect a true pope who alone would have the power to absolve
those in Rome. They did agree to obligate the next pope to work for unity
and be ready to resign, should the majority of cardinals deem it appropriate.
Once the balloting began, it became clear that neither the twelve French
cardinals nor the six Italians saw much gain for themselves in being elected.
They settled on Pedro de Luna, the senior of the two Spanish cardinals present, who had been a cardinal at the infamous conclave of 1378, one of four
surviving participants. His was the only vote against his election. After refusing the office for a time, he agreed to become Pope Benedict XIII, to
honor the previous Benedict, who had been a reformer.

The schism continued. The best opportunity to heal it, the deaths of the
two rival popes, had been allowed to pass by. Benedict had spoken against
cession in conclave but had sworn the oath when the majority of cardinals
required it. Now that he was pope of half of the Church, he declared that
nothing could impede his authority, including any oath that he had taken in
conclave. Neither pope was willing to use the way of cession. The way of action also appeared ineffective, given the balance of power between the two
factions. The way of the council appeared more and more as the only possible solution. The University of Paris in particular supported that approach
to ending the schism. Among its theologians were radicals who declared:

"Of what importance is it if there are two or three [popes], or ten or twelve? Each nation ought to have its own pope."[1] Benedict was not eager for a council, and he proposed that the two popes meet along with their cardinals and hammer out a solution. Any chance that Boniface IX might have accepted the idea disappeared when Benedict found himself in serious difficulty with the French king, and Boniface gained hope of winning the obedience of his rival's major supporter. Benedict's refusal to consider resigning was not well received in France, nor was his lack of generosity to French prelates and princes. In 1398 Charles VI withdrew from obedience to Avignon, although he did not transfer it to Rome. In the fall, French troops placed Benedict under siege in Avignon. It lasted until early 1403, when Benedict slipped away from the city, vowing that he would never again set foot in it.

Meanwhile, Boniface IX was prospering well enough in Rome. Building an efficient bureaucracy permitted him to increase his revenues beyond what Gregory XI had been collecting from the whole Church. He died in 1404, not yet having turned fifty. Benedict's agent was in Rome at the time trying to convince Boniface to meet with the Avignon pope and resolve the schism face to face. Boniface's nephew slapped him into prison at his uncle's death, presumably to prevent him from negotiating with the Roman cardinals. The cardinals entered conclave on the tenth day after the pope's death and proceeded to elect another native of Naples, who became Innocent VII.

After sixteen years of popes living peaceably in Rome, the longest period in 200 years, Innocent found himself the target of a rebellion in 1405 caused by the outrageous behavior of his nephew. He fled to Viterbo. At that time Benedict XIII was in Genoa, where he had gone in hope of meeting Innocent somewhere in Italy and settling the schism face to face. If Benedict had then made a dash for Rome and taken control of the city, as he had a good chance of doing, he probably would have ended the schism in his favor. But he lacked the boldness to take the chance, and the opportunity slipped away. Innocent returned to Rome and died there in 1406. This time his cardinals waited for the news to reach France in hopes that the French king would now decide to abandon the Avignon pope. The time lag in communication between Paris and Rome was too long, however, as the Roman cardinals felt after a month that they had waited long enough and went ahead with their conclave. A promising message from France arrived only after the election. An elderly cardinal from Venice was chosen as Pope Gregory XII. He swore to the cardinals that he would do everything possible to end the schism. He and Benedict agreed to meet, but bad weather and doubts about the other's good faith prevented any meeting, although at one point they were only forty miles apart in northern Italy.

Cardinals from both sides did meet at Pisa in early 1409. What began as a private discussion on how to persuade the two popes to resign soon grew into a general council. The two popes were summoned to appear, but neither did. In June 1409, the assembled prelates declared that both popes were deposed for their many sins and crimes, and they were burned in effigy. The cardinals present, who now totaled twenty-six, went into conclave in Pisa's episcopal palace. It was as free from outside intervention as any conclave yet. After eleven days they unanimously elected a seventy-one-year-old Franciscan friar whom Innocent VII had made a cardinal. Greek by birth but raised in Italy, he took the name Alexander V. He had been an early supporter of the council, which made him acceptable to the Avignon cardinals. Both Gregory and Benedict refused to recognize the authority of the Council of Pisa to force them out of office. Gregory proved to be the greater loser in that far more of his supporters were represented at Pisa; he had little of a curia or a treasury left. Alexander sent legates across Europe with the good news that the schism was over. It was received with joy in most places, and for a short time it appeared that it was true.

Yet, enough princes and churchmen remained loyal to Benedict to make dubious the claim the schism had ended. Ladislas of Naples, who controlled Rome, refused to allow Alexander to enter the city, so he could not take possession of the papal base. Only ten months later, in May 1410, the Pisan pope died at Bologna. Seventeen cardinals gathered for the conclave and after three days unanimously elected Baldassare Cossa, a native of Naples. He had proven his ability as a pirate, military commander, and politician, and as an administrator both before and after being made a cardinal by Boniface IX. A story has it that in the conclave he demanded that the stole of St. Peter, the first symbol of papal authority given to a new pope, be brought to him. He then said to the other cardinals, "I will place it on the man who deserves most to be pope!" and put it around his own neck.[2] He called himself John XXIII. Energetic and capable, he seemed the right man for the task of restoring the unity of the Church under the Pisan papacy. He soon set out for Rome and made good his entry. Yet his luck turned bad nearly as fast. Ladislas of Naples, recently driven out from Rome, returned and expelled John in 1413. Charles VI of France returned to obedience to Benedict at the same time.

John, not one to be stymied by reversals, allied with Holy Roman Emperor Sigismund. The emperor, seeing that the Council of Pisa and its popes had not healed the schism, which was obvious by 1413, ordered his papal ally to call a general council to meet in the Empire. John agreed to convene a council in the south German city of Constance for 1414. In terms of the number of persons present for a council, the Council of Constance was the largest in history; one official counted 72,000 outsiders in the city in 1415,[3] although the Fourth Lateran had more clerics present. Its first order of business was the schism. Expecting that his two rivals would again be condemned, John was most surprised when he was told that the council and the emperor were requiring him to resign because of his misdeeds. After a delay he agreed to resign effective when the other two popes did.

Recognizing that Benedict XIII certainly would not resign even if Gregory XII did, the council first deposed John for his many crimes on May 29, 1415. He gave in and abdicated a week later. The council then turned its attention to Gregory, who was regarded as the more likely to agree to give up. In exchange for abdicating, the council promised to restore him as a cardinal and give him several lucrative church positions. Already over eighty and tired of it all, Gregory agreed, although he refused to abdicate in person before the council, as its members wanted. He formally convened the council, thereby giving it legitimacy in the eyes of those who regarded him as the true pope. Since it is his line of popes that has been accepted by the Vatican as legitimate, the official position is that his act made it a true church council from that point on, unlike the Council of Pisa, which is not so recognized.

Benedict XIII, now residing in Spain, made his own proposal to the council. He would give up his claim to be pope on the condition that all cardinals except those who had the office in 1378 also resign, since their claims to be cardinals were even more dubious. The survivors from that infamous conclave would then elect a valid pope. Benedict was in fact the only cardinal from 1378. Benedict thus claimed the right to choose the next pope by himself. He promised that he would not vote for himself if the council accepted his plan. No one was ready for that. The council declared that the deposition done at Pisa was still in effect, since Constance regarded itself as the successor to Pisa, and set about to elect a pope.

There were in Constance twenty-three cardinals out of thirty-one men who had a claim to the title. The council, not trusting them to make a decision in the best interests of the Church, after long debate added thirty prelates, six from each of the five nations (English, French, German, Italian, Spanish) deemed to be present in the council. Two-thirds of each group had to agree on electing the pope. The fifty-three men, the largest number yet for a conclave, were locked in the merchants' warehouse that the city converted for their use on November 8, 1417, after swearing an oath before Sigismund that they would choose a worthy pope. On the day of entry into the conclave, the solemn Mass for electing the pope, the sermon on electing the best man, and the singing of *Veni Creator* took place. The doors were bricked up and the windows boarded up so that neither light nor air could enter the place. Each elector had a small chamber with a bed, and a table large enough for two. A smaller chamber was built for the conclavists. A wall was set around the building, and a guard made up of nobles from all nations was placed on the wall and the lake alongside it to allow no one closer than the distance of a crossbow shot. The emperor prohibited the people from plundering the possessions of the pope-elect, "even though it were a custom to break into the home of the man chosen pope and loot it of everything it contained."[4] No card games or games of any sort were permitted in the city while the conclave continued.

The electors were obliged to report the results daily to the council; otherwise the conclave adhered as closely as possible to the rules of conclave. It was already a practice for electors to put more than one name on their ballots, but they varied in how they made out their ballots. Some wrote their first choice and then indicated alternatives; others simply gave a list of names; still others only wrote one name. On the first day, there were six candidates with some strength; on the second, Odonne Colonna received fifteen written votes from cardinals, one short of two-thirds, and more than enough from the electors from the nations. The procedure of accession then followed. That was the practice following a scrutiny in which the electors would be asked if anyone wanted to change his vote to one of the stronger candidates in the hope that this would complete the election. In this era it was done verbally. For some time no one spoke up. Then two cardinals, who had been conversing together and were said to dread voting for Colonna, decided that someone was certain to accede to him. So they announced: "For the consummation of this work and the union of the Church we two transfer our votes to the lord Cardinal Colonna."[5] All others then acceded to him as well. He became Pope Martin V to honor the saint on whose feast he was elected. He was a cardinal in the line of the Roman papacy but had helped elect Alexander V at Pisa. The Colonna family had long been a power in Rome and the Church, having had twenty-seven cardinals to this point, but Martin was its only pope. If one can argue that the beginning of over a century of troubles for the papacy was Boniface's excommunication of the two Colonna cardinals, then it is fitting that it would end with the election of a Colonna pope. He was still a subdeacon, the last one to become pope, so in short order he was ordained deacon, priest, and bishop before being crowned pope.

As for the other popes, Gregory XII was already dead. John XXIII acknowledged Martin as pope and received his reward by being restored as a cardinal. He died in 1418 at Florence and was buried in a magnificent tomb in its cathedral. When word reached Benedict in Spain that he had been deposed and a new pope elected, he excommunicated the entire council, but his three cardinals met in their own conclave in 1418 and voted for Martin. Scotland officially remained in obedience to Benedict for another year. After that he was reduced to a small handful of loyalists at his court in Aragon. The realm of Aragon formally was not in obedience, but its king protected him. He died in 1423, age about ninety, soon after he named four new cardinals. His was the longest reign—twenty-nine years—of anyone who had a claim to be pope until Pius IX, but the fact that he had lived beyond "Peter's years" was used against him in his later condemnation as antipope. Holding their own conclave in the room below the one where he died, his cardinals elected a canon lawyer as Clement VIII. In 1429 he submitted to Martin. A diehard cardinal holed up in an inaccessible fortress in the Pyrenees then elected a French prelate as Benedict XIV. Little is known about his life except that he died in 1447, but in 1469 a French blacksmith was burned for declaring that Benedict had ordained him a priest years before. Thus ended the Great Schism of the West.

For several reasons the popes of the Roman line have been accepted on the official list of popes as the rightful popes to come out of the election of 1378. One is that their presence in Rome gave them the high ground. A second is that Urban had been elected first, and while his election was irregular, it has not been seen as so defective as to be invalid. Third, Catherine of Siena, whose opinion counted for much in this matter, acknowledged him as pope, although she also criticized him for his violent behavior. Perhaps the most important reason is that when a solution was finally found in 1417, the Roman pope submitted and resigned, while the Avignon pope remained obstinate and continued the Great Schism. The papacy has demonstrated that it holds Urban VI and his line of popes to form the true succession from St. Peter by reusing the numbers of the Avignon and Pisan popes, most recently with John XXIII, except for the next Alexander, who became Alexander VI. That, along with the long delay until another John XXIII came along, suggests that the papacy has been more ambiguous about the standing of the Pisan popes than those at Avignon. Despite the papacy's later judgment, the issue of which was the rightful line to come out of the election of 1378 lies too evenly balanced for the historian to give a definitive answer. In 1415 John held the obedience of a far greater part of the Catholic population than did the Roman pope.

Although Sigismund tried to make Martin V reside in Germany, Martin went to Rome in 1420, finding the city in worse condition than it had been in 1377. Still he remained there. If there was a positive result of the schism, it lay in the fact that the popes never again would voluntarily take up long-term residence outside of Rome, although they would often be elsewhere for months at a time. With Martin began also a powerful tradition that the pope would be Italian. Of the fifty-seven popes from Martin to the present, only four have not been Italian.

In respect to the matter that sparked the schism, the election of the pope, little was changed. Despite the significant innovation at Constance of having electors from the Catholic nations, Martin's legitimacy went unchallenged except for those few who still recognized Benedict. That innovation disappeared immediately; no one lamented its demise. Since 1417 the conclave has functioned well in that the legitimacy of the popes chosen through it has been unquestioned except by the most minuscule numbers of Catholics, even when those choices keenly disappointed powerful Catholic monarchs. That record is unparalleled in the history of institutions.

The council limited the number of cardinals to twenty-four. Since Martin had bought recognition by accepting the cardinals of the three rival popes, it was only in 1426 that the number dropped to that number. Martin's cardinals reveal one result of the schism. He made the College a more international body by appointing the first German cardinal in over two centuries along with two Spaniards, an Englishman, and a Cypriot, as well as the usual French and Italians. When he died in 1431 at Rome, nineteen cardinals assembled in the Dominican house of Santa Maria sopra Minerva, since the small number did not require a large space for the conclave. Domenico Capranica, who had been publicly proclaimed a cardinal but had not received the red hat because he had been away from Rome on papal business, was not admitted to the conclave. He was allied with the Colonna, and the College's bad relationship with Martin sparked the decision to exclude him to reduce the Colonna influence in the election.

Angered by Martin's efforts to reduce their incomes and privileges, the cardinals produced a capitulation to limit the pope's authority over the College. There had been earlier conclave attempts to influence papal policy, but this was the first formal document intended to bind a new pope. Its most significant clauses required the pope to share half of the papal revenues with them and not to decide any major issues without their consent. The cardinal elected, who took the name Eugene IV, had sworn to the document and, on becoming pope, issued the necessary bulls to put it into effect. But he withdrew the concessions after his relationship with the College became highly difficult.

The new pope was the nephew of Gregory XII, who had made him a cardinal in 1408. He had been active at Constance and was a recognized leader among the cardinals in supporting the restrictions on papal power. As so often has happened in the history of institutions, and certainly of the papacy, one who has been eager to reduce the power of the leadership position in an institution believes differently once he gains that position. The key issue for Eugene was Constance's decree that required a general council to meet on a regular basis—the first one in five years, the next after seven, and every ten years from then on. Martin had held a council at Pavia in 1423, but it was so poorly attended and did so little that it is not counted as a general council. In 1431 he called a council for Basel, but his ill health delayed its opening until after his death. Eugene was faced with dealing with it. When the emperor invited representatives of the Hussites in Bohemia, who had been denounced as heretics at Constance, the pope ordered the council dissolved. The reaction against Eugene was fierce, and he was forced to back down, but he became a dedicated foe of the council. In 1439 he ordered its members, who had been meeting off and on at Basel, to move to Florence. A large minority, mostly Italians, obeyed. At Florence they met with Byzantine Emperor John VIII, who was prepared to accept reunion with Rome in exchange for Western help against the Ottoman Turks. Most Orthodox Christians rejected his acceptance of reunion, and the limited aid sent to Constantinople failed to prevent its conquest by the Turks in 1453.

The council members at Basel were outraged by Eugene's attempt to break up their meeting. In November 1439, they voted to depose him and established a electoral body of thirty-two churchmen to enter conclave. They elected the former duke of Savoy, who had recently abdicated his duchy, as Pope Felix V. It appeared as if the schism was about to begin again. Felix's election, however, was far more dubious than Clement VII's in 1378, since only one cardinal was present. Only the diehard conciliarists, those who affirmed the theory that the council held authority over the papacy, accepted him as pope. When Eugene excommunicated Felix and the council, Eugene was in Florence, where he had fled in 1434 after a revolt in Rome led by the Colonna. They resented his reduction of the privileges and power that Martin had given them. It was at least the twenty-sixth time a pope fled

Rome because of violence directed against him, but it was the last until 1800.[6] Eugene remained in Florence for nine years, until the situation in Rome turned in his favor. There he died in February 1447.

Eighteen of the twenty-four cardinals, all but two created by Eugene, entered conclave on March 4, 1447, again in Santa Maria sopra Minerva. In order to gain the support of the Catholic rulers against the council, Eugene had internationalized the College—four Spaniards, two French, and one each from England, Germany, Hungary, Poland, and Portugal. Most of them were what later would be termed crown cardinals—men chosen for their service to their kings, not to the Church or the papacy. Five non-Italians were not present, so the eleven Italians controlled the voting, but they were divided again between Colonna and Orsini. Prospero Colonna had ten votes at the first scrutiny, but he failed to pick up two more by accession for election. In the second he had the same number but also held the promise from three cardinals to give him their accession if it was the last vote he needed. No one spoke up to give him the eleventh vote. In a powerful demonstration of how an unexpected candidate can be elected when the favorite stumbles, on the third ballot Tommaso Parentucelli, the Cardinal of Bologna, who had been a cardinal for only three months, received the two-thirds majority required. The result was so surprising that the scrutators did a second count to be sure. The rest of the cardinals agreed to accept him, so the election was announced as unanimous. He took the name Nicholas V. He had a degree in theology and was one of the few popes at that point with a university degree; the others who did were canon lawyers.

Felix V sent Nicholas a bull ordering him to come to Basel to answer charges of perpetuating a schism. Nicholas, of course, did not go, but he used gentle words and offers of church positions to lure Felix's supporters away from him. When a half-year later Emperor Frederick III abandoned Felix and the Council, there was no hope of making good their claims to authority in the Church. With the help of French King Charles VII, an agreement was drawn up in 1449 by which Felix renounced his claim to the papacy and Nicholas absolved him and his followers of excommunication. The conciliarists, on the pretense that a vacancy existed in the Holy See, elected Nicholas as pope, thereby maintaining the claim that the council had authority over popes. Felix was made a cardinal and died in 1451 without ever going to Rome. He was the last antipope to be elected by at least one cardinal and the last to have any support in the Church. The shadowy figures who have claimed to be pope since then have had no support beyond a small circle of believers. Schism did not disappear from the Catholic Church, but in the future it would no longer affect its head. Felix's submission marked the end of conciliarism as a viable theory of church governance, although it remained strong in France and would be taken up by the Protestants.

Nicholas survived a plot in 1453 against papal rule of Rome by the humanist Stefano Porcaro, who intended to expel him from Rome or assassinate him if necessary. There is no doubt that Porcaro wanted to create a Roman republic free of the papacy. The suppression of the revolt and Porcaro's hanging demonstrated that papal authority in Rome was becoming more effective. Nicholas's disappointment in Porcaro, whom he had treated generously before the conspiracy, sapped his health, and he died a year later lamenting the burden of the papal office. He complained that the Cardinal of Bologna saw more friends in a day than Pope Nicholas would see in a year. Ten days later, on April 4, 1454, fifteen cardinals entered conclave at the Vatican palace, which would serve as the site for all but six conclaves to the present.

With its location now consistently in the Vatican, the conclave took on the characteristics that were standard for centuries. The cells in the Vatican were about twenty-five square feet in size and were so dark that artificial light was required in nearly all of them. The cells were marked with letters, and their occupants chosen by lot. The conclave was kept by three sets of guards—ambassadors, Roman citizens, and prelates. Each set checked items that were passed into the conclave to ensure they did not carry any message from the outside; for example, they poked into the food that a cardinal's servants brought to the *rota* (revolving doorway) for their master's meals.

By this time the role of the conclavist had become defined. (The second man whom cardinals were permitted to bring in was usually a servant.) He was more than just a secretary for the cardinal; he was his confidant and often his contact with other cardinals, usually via their conclavists. Many cardinals preferred to have conclavists negotiate the promises of favors to the others, which played so major a role in papal elections for the next three centuries. The conclavists did much of the bargaining attributed to the cardinals, often in clandestine meetings that determined the outcome of conclaves. Talented conclavists could achieve a great deal for their masters, but because so much of what they did was behind the scenes, it is difficult to assess accurately their place in the elections, although many conclavists wrote diaries and memoirs detailing the events of the conclaves and their roles. Brothers or nephews of cardinals often served as conclavists until Pius IV banned them. Surprisingly only three popes were conclavists earlier in their careers.

Six cardinals were absent from Rome when Nicholas V died, but one made it back in time to be admitted. Several of those absent were Italian, so there were only seven Italian cardinals in the conclave. The next largest contingent was the Spanish, with four members. Again the key division in the conclave was between Colonna and Orsini. The Colonna could count on nine votes for Prospero Colonna but had no hope of picking up one more to make the required majority. The Orsini backed the respected cardinal Pietro Barbo. When it became clear after two ballots that neither would be elected, sentiment swung to the Greek churchman John Bessarion, who had come to the Council of Florence in 1439 and stayed on, becoming Catholic and a cardinal the same year. At least eight cardinals regarded his learning, piety, and talent as worthy of the papacy. His election appeared to be so certain that several cardinals asked favors from him. A French cardinal, determined to stop his election, made the rounds of the others, raising questions about Bessarion's faith and sincerity; "It is a only short time since he attacked the faith of the Church of Rome. . . . Should he now be our master?"[7] The boomlet for Bessarion eroded, and the cardinals were back to square one.

After three days of unfruitful balloting, the Roman crowd and the ambassadors of the Catholic states began to make clear their dissatisfaction over the failure to elect a pope. Several noncardinals were proposed but rejected. Finally the cardinals turned to the eldest among them, the Spaniard Alfonso Borgia. He was elected by accession. He chose the name Callistus III. At seventy-seven years old, he was chosen as a caretaker pope, and he fulfilled that purpose reasonably well by reigning for just over three years. The Italians were disappointed in his election, and their frustration with him increased as he gave out patronage to fellow Spaniards. Several Italian cardinals left Rome amid talk that they would create a new schism. Callistus died in August 1458 before matters reached the point of a schism. He made his mark in history largely by his extreme nepotism. His many relatives swarmed to Rome to seek high positions in the Church and the city. Two nephews were erected as cardinals, and one complaint made about them was that together they still did not have enough years for a red hat. The more notorious was Rodrigo, created a cardinal in 1457.

Another man Callistus made a cardinal was Aeneas Sylvius Piccolimini, a noted humanist and skilled diplomat, who was called the Cardinal of Siena. His memoirs provide the most extraordinary source for the history of papal elections. He had taken part in the two previous conclaves as a conclavist. Now in 1458 he was eligible to vote in the conclave that opened ten days after Callistus's death. There were nineteen cardinals close enough to Rome to participate, but the Cardinal of Fermo, whom everyone expected to be elected, died just before entering the conclave.

By then the word *prattiche* was used to refer to the discussions that took place before the conclave began about the candidates and strategies for electing them. Also by that time, several Italian states had established professional diplomatic services, and the reports of the ambassadors in Rome from those states are highly informative on the conclaves. Francesco Sforza, duke of Milan, had strongly supported the Cardinal of Fermo, telling his ambassador to use every means to ensure his election. After Fermo's death, the Milanese ambassador wrote to Sforza indicating that he thought Aeneas could be elected, whom he saw as favorable to the duke.

The eighteen cardinals included eight Italians, five Spaniards, two Frenchmen, two Greeks, and one Portuguese. This conclave was the last one until 1958 in which non-Italians constituted a majority of the voters. It took place in the Vatican. A large chapel was transformed into cells for the cardinals, while a smaller one was used for the scrutinies. On the first day, no balloting was done. The cardinals worked on a capitulation containing the restrictions on papal power that the cardinals wanted to impose on the new pope. It mandated seeing to the financial well-being of the poorer cardinals and required the new pope to mount a crusade against the Ottoman Turks.

It was only on the morning of the third day that the first ballot took place. Some put as many as five names on their ballots, and Aeneas commented sourly that Cardinal Orsini had seven names on his ballot, so he would be sure to vote for the winner. The tally of the first ballot revealed that Aeneas and another cardinal had five votes apiece, while no one else had more than two. None was cast for Guillaume d'Estouteville, the Cardinal of Rouen, who was known to be ambitious for the office. At this point there should have been an opportunity to elect by accession. The Cardinal of Rouen and two other powerful cardinals objected to this, because they had received no votes and were therefore ineligible in this process. The cardinals agreed to abandon the practice for the remainder of the conclave. Then they broke to eat, and, as Aeneas put it, "the richer and more powerful members of the college . . . begged, promised, threatened, and some, shamelessly casting aside all decency, pleaded their own cause and claimed the papacy as their right. . . . Their rivalry was extraordinary, their energy unbounded. They took no rest by day or sleep by night."[8]

Aeneas alleged that Rouen recognized him as the greatest danger to his own election and put great effort into slandering him. Rouen claimed that Aeneas wanted to transfer the Apostolic see to Germany, where he had served as a diplomat, and that he was a pagan because of his enthusiasm for Roman poetry. Rouen also made extensive promises about the church positions and wealth that would come to those who voted for him. His efforts paid off, as other cardinals began to back his candidacy, meeting in the privies, "as being a secluded and retired place" to engage in their bargaining. Aeneas commented, "Where could one more appropriately enter into a foul covenant than in privies!"

The memoirs indicate that there was only one scrutiny each day. The overnight efforts of Rouen and his friends appeared to have been successful, since they believed that Rouen would get eleven votes and needed only one more. "They did not doubt that they would at once get the twelfth. For when it has come to this point, some one is always at hand to say, 'I too make you pope,' to win the favor that utterance always brings." A friend of Aeneas woke him early with the news that the election was as good as done. He urged the humanist to go quickly to Rouen, as he himself was going to do, and tell him he was going to vote for him, out of fear that if Rouen was elected without him, he would make trouble for Aeneas. The cardinal commented that this had happened to him in respect to Callistus, who never forgave him.

Aeneas responded with a denunciation of simony (embellished after the fact, one suspects). He said that if Rouen would force him to live in poverty, the Muses "are all the sweeter in poverty." Said Aeneas: "Tomorrow will show that the Bishop of Rome is chosen by God, not by men." He credited his speech with persuading his friend not to vote for Rouen. He went to several other cardinals exhorting them with strong words about the dangers of electing someone who would transfer the Holy See back to France and make Italians slaves of the French. When an Italian objected that he had pledged to vote for Rouen and could not break his word, Aeneas thundered: "You now have to choose whether you prefer to betray Italy and the Church or the Bishop of Rouen."

In the balloting that morning, Rouen was one of the scrutators who counted the votes. Aeneas and his allies feared that he might try to cheat, so they carefully wrote down each name as it was read off the ballot. When Rouen announced that Aeneas had received eight votes, he objected that his own tally showed nine. A recount verified Aeanas's charge, to Rouen's chagrin. The cardinals, Aeneas claimed, were amazed at the voting, because never in anyone's memory had any candidate received so many votes through the written ballot. Despite having agreed earlier not to use the method of accession, they decided to invoke it in hopes that they could complete the conclave then and there. For some time everyone sat silently. Then Rodrigo Borgia, the youngest cardinal, announced that he was changing his vote to Aeneas. After a longer pause, another cardinal did so, making Aeneas one vote short of victory. Colonna then rose with the intention of putting him over the top. Rouen and a friend physically tried to restrain him from speaking but could not prevent him from shouting: "I accede to the Cardinal of Siena and I make him pope." All the cardinals then came to kiss his feet. Bessarion, who had voted for Rouen, said that he greatly admired Aeneas's virtues but had believed his health was too poor for the burden of the papacy. Aeneas made a conciliatory reply and then announced that he would be called Pius, from "pius Aeneas" of Virgil's *Aenied*. It was then announced from a high window to the crowd outside that the Cardinal of Siena had become Pope Pius II.

The new pope's memoirs also described how the Roman mob, hearing that he was the new pope, went to plunder his house, taking everything, even the marble blocks, while the conclave attendants ransacked his conclave cell. The practice would later be interpreted as signifying the end of the new pope's old life. The houses of other cardinals also suffered plundering, as rumors swept the crowd that various cardinals had been elected. The Cardinal of Genoa suffered that fate because his name was confused with Siena when the announcement was made. Pius II went on to say how happy everyone was that he had been elected, not only the Roman people but also most Catholics, albeit not the French or the Venetians. He commented that he was almost killed when the Roman mob fought over who was to get the horse he rode during the procession to the Lateran. Pius's description of his election is by far the most extensive to come from a pope. He surely embellished the dramatic tale of his election, but there is no reason to suppose that he misled in his description of the political machinations that went on in a conclave. It can be assumed that most cardinals and new popes went through much the same range of emotions and skullduggery in the conclaves long before and after this one.

CHAPTER V

Conclaves in the Renaissance

Pius II, elected in 1458, can be called the first Renaissance pope. A humanist himself, he was an eager patron of artists and fellow humanists. He was determined to beautify Rome and make it a city worthy of the papacy; transforming Rome became a major concern for the papacy for the next two centuries. Nepotism was another trait associated with Renaissance popes. Many an earlier pope had given a red hat to a nephew, but Pius and most of his successors for the next 250 years went well beyond that. He made three nephews cardinals, and other relatives were given lucrative positions in the Church and the Papal States.

Pius's primary goal was recovering Constantinople from the Turks. In 1459 he called all the Christian rulers to Mantua to establish a general peace among them and commit to a crusade. It required a long absence from Rome. He issued a bull requiring that the next election be held in Rome regardless of where he might die.[1] Pius believed that conclaves held elsewhere led to schism. While his congress was well attended, it failed in its goal, and he returned to Rome in 1460 after a year's absence. He did not abandon his objective, and he died in August 1464 at an Adriatic seaport while planning an expedition against the Turks. The twenty cardinals obeyed Pius's order to bury him in Rome and hold the conclave there. It took fourteen days before it began in the Vatican. By 1464 the diplomatic correspondence is extensive enough that there is good information on who were the *papabili* (popeables), a term by then in use. Cardinal Bessarion was again the favorite.

One ambassador worried that so many cardinals wanted to be pope, they would deadlock the conclave and force a noncardinal's election. The international College of previous decades had returned to being a majority Italian. Despite fears that Pius II would be too favorable to the Germans, his cardinals were all Italians. His three nephews were too young to form a powerful bloc in the conclave, however.

The first order of business for the cardinals was drawing up a capitulation, which one refused to swear to. The many provisions of this capitulation, the longest one yet, included the standard demands of keeping the College at twenty-four members and requiring its approval for new ones. Only one member of the pope's family could be admitted into the College, and the minimum age for a cardinal's hat was set at thirty years. No member of the pope's family could serve as commander of the major fortress in the Papal States or hold any of the major political offices; this was directed at the dead pope, who had filled those positions with his relatives. The pope was obliged to call a council in three years, and there was a long list of other checks on papal freedom. Twice a year the cardinals would determine how well the pope was observing the clauses of the capitulation. They were to admonish him three times if they determined that he was failing to live up to his promises. Left unsaid was what recourse the cardinals had should the pope refuse to heed their admonitions.

Once the cardinals turned to balloting, they reached a conclusion immediately. Pietro Barbo received eleven votes in the first scrutiny on August 30, 1464. He then picked up three more by accession to reach the required number, and the rest saluted him as pope without hesitation. He announced his intention to become Formosus II, but it was objected that the word means handsome in Latin, and Barbo was known for his vanity about his appearance. He became Paul II. This is the only time that a new pope had to settle on a second name. The Roman authorities had taken precautions to prevent the mob from plundering the new pope's house in the city. When the mob arrived at the house, it was well enough guarded to prevent it from being ransacked, but it was only after 1,300 ducats were distributed that the mob went home. Thereafter, new popes distributed large sums of money in hope that their houses would not be plundered, but it took three centuries before the practice of ransacking the new pope's house disappeared entirely. The conclavists for their part took the furnishings of the new pope's cell and also expected to receive money and benefices from him to compensate them for the inconvenience of being in the conclave. In 1572 Gregory XIII refused to give them anything because that conclave lasted only one day.

Paul was the forty-eight-year-old nephew of Eugene IV, who had given him a red hat in 1440. Through his mother he was a grand nephew to Gregory XII. Paul loved pomp, and his coronation reached heights of splendor and spending not seen in Rome for centuries. Once enthroned, he denounced the election capitulation, which limited a pope to appointing only one of his relatives to the College. Paul named three nephews. He gave red hats to sixteen men, all but three Italians. Several soon died, so he seems not to have ever put the College over twenty-four members. Paul was concerned with avoiding a council, calling a crusade against the Ottoman Turks, and keeping the peace among the Italian states. Despite his conclave pledge, he did not call a council. He made a valiant effort to organize a crusade against the Turks, but bitter rivalries among the Christian states prevented a force from being assembled. Paul was probably the most successful pope of the fifteenth century in maintaining the peace across Italy. A minor foible is blamed for his death in July 1471: He overindulged in chilled melon on a hot afternoon in Rome. By midnight he was dead. That his sudden death did not give rise to rumors of poisoning suggests that nobody believed that anyone wanted him dead.

Quickly eighteen cardinals gathered in Rome. Two of Paul's appointments were not allowed to participate because their promotions had not been published, despite his order that it be done immediately at his death. Those who entered the Vatican included fifteen Italians and three from outside Italy, a dramatic change from 1447 when non-Italians were the majority. Bessarion, d'Estouteville, and Filippo Calandrini were the *papabili*. The latter two waged aggressive campaigns for votes. The first order of business, however, was the capitulation, which was little different from previous ones. Bessarion, regarded as the strongest candidate, hurt his chances by refusing to swear to the capitulation, which gave the others the impression that as pope he would not consider their needs. A story also has it that three cardinals wanted to talk to him about the favors he would do for them if they voted for him. His conclavist turned them away, so they approached Francesco della Rovere, who was far more amenable. Supported by the Sforza of Milan, he emerged as the new pope.[2] It is not certain that it occurred in the third scrutiny, but the conclave ended on the third day. Della Rovere took the name Sixtus IV to honor St. Sixtus II, on whose feast day the conclave had begun.

The new pope was born into a fisherman's family near Genoa. Joining the Franciscans, he was their general when Paul II named him a cardinal in 1467. He was known as a learned theologian and devoted follower of St. Francis, and the more spiritual-minded Catholics were pleased with his election. Unfortunately, the worldly cardinals who voted for him had a better gauge of his character. Upon becoming pope he provided the cardinals with rich gifts and offices to pay off his promises in the conclave. Ignoring the capitulation he had sworn, he gave red hats to six nephews, including Pietro Riario, rumored to be his son. One of them, Giuliano della Rovere, became famous as Julius II, but Girolamo Riario, a layman, garnered more attention in Sixtus's reign, most of it bad. Sixtus began to carve out a territory in the Papal States as a hereditary principality for Riario. The objections of Florence, close to the lands designated for Riario, led to papal involvement in the Pazzi Conspiracy against the Medici in 1478. Sixtus excommunicated the Medici and placed Florence under interdict in response to the archbishop of Pisa's execution after the conspiracy failed.

Sixtus's policies created great resentment across Europe, and Andreas Zamometic, a Slavic bishop and diplomat for Emperor Frederick III, thought the moment right to convoke a council to reform the Church and depose the pope for his misrule. At Basel in March 1482 he called a council into existence, but the time for conciliarism had passed, and no one showed up. Catholics were becoming used to the behavior of the Renaissance papacy exemplified by Sixtus IV. His appointments to the College were intended to eliminate opposition to him from both the cardinals and the major Italian city-states by naming members of the ruling families. Many of his cardinals were not priests, and they introduced an element of worldliness well beyond that of previous eras. He essentially created the Renaissance College. The growing discrepancy between those cardinals who had enormous private

wealth and those who did not became a major problem, as the poor cardinals tried to keep up with the extravagant lifestyles of the wealthy. The pope
was obliged to increase the poorer ones' incomes, which put a great burden
on papal revenues and led to the financial exactions on Catholic countries,
especially Germany, that helped spark the Reformation. The need for more
money led to accepting bribes from kings and wealthy city-states wishing to
influence the conclaves. Certainly, favors and money had changed hands before, but now it became standard for great sums to be dispensed before and
during an election. Sixtus largely was responsible for another key aspect of
the Renaissance papacy—establishing his family as a princely house in Italy
by appointing relatives to wealthy church offices and marrying them into
noble families. Yet he also began the extensive building program that created
Renaissance Rome.

When Sixtus died in August 1484 in Rome, Girolamo Riario was away
campaigning with the Orsini against the Colonna. Caterina Sforza, his wife,
seized control of the fortress Castel Sant'Angelo, and it took two weeks to
negotiate her surrender and ensure that the conclave was reasonably free.
The Romans hated Sixtus and plundered the properties of his relatives and
the Genoese. Riario's hope that he could control the conclave was abandoned in view of Roman hostility toward him. Because of the violence in
Rome, many cardinals did not attend Sixtus's funeral, but all twenty-five in
the College assembled for the conclave on August 26. Only four were not
Italians. Sixtus gave red hats to thirty-four men, but many, including three
nephews, had died. His appointments made up a majority of the electors, allowing Giuliano della Rovere to be the conclave's power broker. The
nephews of Pius II and Paul II were regarded as the strongest candidates,
while another papal nephew, Rodrigo Borgia, was actively campaigning for
himself. Because of the many cardinals, the factions were more numerous
than in recent elections. Observers expected a long and difficult conclave.

The conclave took place for the first time in Sixtus's new chapel in the
Vatican. It was large enough to accommodate the cells of the twenty-five cardinals and their conclavists as well as space for voting. Each cardinal had a
small bedroom and a sitting room. The two conclavists had small rooms
above that. The cells were airless and dark except for a few along the outside

walls where the windows were left open at the top for a little light and air. When the number of cardinals in future conclaves passed thirty, cells spilt out into adjoining halls, and in 1565 the Sistine Chapel was vacated as dormitory space and used exclusively for balloting. Before Michelangelo's famous paintings in the chapel were done, there already were Perugino's *Christ giving the Keys to St. Peter* and Signorelli's *Moses giving the Rod to Joshua*. Both were regarded as auspicious for the cardinals who had the good fortune of getting the cells under them. One cell also deemed lucky was constructed on a step leading to the papal throne, which raised that cell above the others.[3] By this time the practice was in place of using purple to decorate the cells and thrones of the cardinals created by the dead pope, while those from prior popes were done in green.

The cardinals first drew up a capitulation, which contained several new points. The most telling was a clause protecting any cardinal whom a secular ruler might seek to punish for the way he voted in the conclave. Emphasis was given to the clauses allowing a pope to give a red hat to only one nephew and keeping the size of the College at twenty-four members. When the cardinals turned their attention to the balloting, they decided to eliminate the process of accession. Jacob Burchard, the papal master of ceremonies, whose diary is a valuable source for four conclaves, related that the reason was the fear that Cardinal Marco Barbo, Paul II's nephew and a staunch reformer, might be elected pope to the detriment of more worldly cardinals.[4] Barbo did get the most votes in the first two scrutinies but was well short of the needed two-thirds. The negotiating then began in earnest among the popemakers.

Della Rovere proved to be the most adept at it. He first approached Barbo and promised to get him the last necessary votes for his election in exchange for several church offices. Barbo rebuffed his offer as simony, thereby showing why many cardinals did not want him as pope. Della Rovere then turned to Giovanni Cibo, a fellow Genoese, who had received six or seven votes in the second scrutiny. He teamed up with Borgia, who had much to fear from a Barbo papacy, and they visited the cells of the younger cardinals at night, not bothering with the older ones. They brought promises of gifts and favors from Cibo if he were to become pope. Some cardinals went to

Cibo's cell to have him sign pledges to nail down what had been promised to them. When della Rovere was sure he had enough votes, he woke up the older cardinals and rushed them still groggy into the chapel. When it became clear that Cibo had the votes needed, the rest of the cardinals acceded to his election. He took the name Innocent VIII, thereby endorsing as legitimate Innocent VII of the Roman line during the Great Schism. There is no proof that he intended to make any such judgment by his choice of a name, but it was in effect what he did.

Della Rovere had chosen well if it was his intention to serve as the power behind the papal throne. Innocent was fifty-two years old but chronically ill and weak-willed. Sixtus IV had made him a cardinal in 1473. He had had two illegitimate children before he became a priest. Everyone with whom he came into contact loved him, but he was hardly the man needed to deal with the viper's nest that was Rome. Rumors of the pope's death filled the next several years. At the spread of the first such rumor a year after his election, the Orsini seized control of the strongholds of the city in preparation for dominating the conclave, only to have to give them up when the pope proved to be still alive. Innocent swung his support to the Colonna, allowing them to recover most of the losses inflicted on them by the previous two popes, whereupon the Orsini called upon the Romans to depose that "Genoese sailor who is not true pope." The Romans refused to be goaded into taking action against Innocent, but in 1486, when another rumor of his death swept the city, they looted many papal properties.

The highlight of Innocent's reign was the marriage of his son to a daughter of Lorenzo de'Medici in November 1487. The pope himself officiated at the marriage ceremony and provided a great feast for the occasion. It was the first time that a pope had celebrated the marriage of his own child, and it set a bad precedent. Part of the reward for providing a bride for the pope's son was a red hat for Lorenzo's thirteen-year-old son Giovanni. (He was not the youngest cardinal ever; a nine-year-old from the eighteenth century has that dubious distinction.) The pope required the lad to wait four years before he was allowed to function as a cardinal. He named only one of his own nephews as a cardinal. Innocent fell terminally ill in early 1492. Soon all he could take was mother's milk. Yet he lived on into the summer, while chaos erupted in Rome. The deathwatch on Innocent was one of the longest for a pope. The *prattiche* was at first ferocious, but by July 25, when Innocent died, calm had settled over Rome.

After the funeral, the first order of business for the cardinals was deciding whether to accept two men into the conclave, whom Innocent had named as cardinals but had not publicly proclaimed. The College agreed to allow them in, thereby increasing the number of those who entered conclave on August 6 to twenty-three. There were eight nephews of previous popes among them. Four cardinals, all non-Italians, were absent. The diplomatic correspondence from long before Innocent's death indicated that Giuliano della Rovere was regarded as the strongest *papabile,* with nine votes certain to be cast for him. Other strong possibilities included Ascanio Sforza, for whom Ludovico Sforza of Milan was campaigning, and Francesco Piccolomini, who was admired for his spiritual qualities. Charles VIII of France was rumored to have put 200,000 ducats into a Roman bank to use for bribes; knowing that a French cardinal could not be elected, he supported della Rovere. Rodrigo Borgia, the senior cardinal in service or dean, was not expected to win, because the mostly Italian College would not elect a non-Italian. He was the only one in the conclave.

After drawing up a capitulation that differed little from the previous conclave's, the cardinals began balloting. In respect to this conclave the problem for historians is not a lack of sources but an overabundance of conflicting ones. Out of the confusing accounts the best scenario that can be constructed for the conclave is that after the first three scrutinies it was clear that della Rovere was slipping, while the other *papabili* had not gained enough. Paul II's nephew had ten votes in the third scrutiny, but that was far from the required sixteen. When at the accession after the ballot he failed to pick up votes, his election became improbable. In the evening after the third ballot, vigorous politicking, or to use less of a euphemism, bribery, took place. Sforza announced that he supported Borgia, who had eight votes in the third scrutiny. He brought with him four cardinals committed to following his lead, so that Borgia could expect to get thirteen votes the next morning. As often happened in the conclaves, once it appeared someone was close to being elected, more cardinals were eager to get on the winning side. Three cardinals declared their intention to vote for Borgia. Then della Rovere did as well, bringing his bloc over. Borgia profusely thanked him for making him pope, but if the above scenario is correct, he already had enough votes. At the scrutiny on the fourth day, the balloting was unanimous for Borgia, who voted for himself.

The standard charge against the conclave of 1492 is that it was the worst example of simony in the history of papal elections. In fact it appears to be less corrupt in that respect than the previous conclave or the next ones. To be sure, Ascanio Sforza emerged with Borgia's office of vice-chancellor and his house in Rome, but as pope Borgia had to give them to someone. Although Borgia had amassed wealth during his forty years as a cardinal, his family was not rich, and the sums he could offer for votes would not have impressed the many cardinals who came from immensely wealthy families. The strongest proof that Borgia had not schemed to become pope by any means lies in the fact that he had not arranged for two Spanish cardinals, one his cousin, to come to Rome while Innocent lay dying for four months. Borgia was an experienced member of the curia who was generally well liked. When the favorites failed to win early on, the cardinals turned their attention to others. If the promise of favors sped that process along, it was the norm for papal elections in that era. It can be argued that Borgia's notoriety as pope led historians of his era and of modern times to project gross impropriety back into the conclave that elected him to explain how so immoral a man could have become pope.

Borgia announced that he would become Alexander VI. Alexander V had been one of the Pisan popes during the Great Schism, but there is no indication that the new pope intended to validate them by his choice. He was sixty years old and the father of six children for certain, including Giovanni, Cesare, and Lucrezia. After becoming pope he was accused of having two more children and taking Julia Farnese, sister of the future Paul III, as his mistress. Fathering children was not unusual for cardinals in the Renaissance, but few had as many as Alexander. He ignored the misdeeds of his children, especially Cesare's, and his effort to establish them at the top of Italian society was a major reason for his notoriety. It is unlikely, however, that he was guilty of incest with Lucrezia, which accusation was used to blacken the reputations of both.

After unknown murderers killed Giovanni in 1497, the pope focused on establishing Cesare as the ruler of a realm carved out of the Papal States. He had made his son a cardinal soon after his election; now Cesare resigned the office on the grounds that his father had coerced him into the clergy. Eager for greater things for Cesare, the pope sent him to France in 1498 to negotiate with Louis XII, who gave him a French duchy and a princess for a bride.[5] In exchange Louis's first marriage was annulled, and his right-hand man, Georges d'Amboise, made a cardinal. Alexander was a prolific creator of cardinals, naming forty-three, which fact is usually attributed to the lavish gifts new cardinals gave the pope. By promoting seventeen Spaniards it is more likely that he was interested in creating a College that Cesare could control in the next conclave. A bloc of old cardinals led by Carafa of Naples was hostile to Alexander despite voting for him. Carafa was ready to support Savonarola's call for a council to depose the pope for simony, but the Florentine's sudden loss of power in 1498 nipped the idea in the bud. At Alexander's death the forty-five cardinals included eight Frenchmen and eleven Spaniards. A bloc of eight French votes in the expanded College could not control the conclave, but Louis XII worked diligently in the years after d'Amboise's creation as cardinal to gain allies for his election. When Alexander and della Rovere had a bitter falling out, della Rovere fled to France. Louis was generous to him, expecting that he would support d'Amboise in the next election.

When the next conclave took place, the French were in a strong position to promote d'Amboise's election. In 1494 Charles VIII had decided to make good the Angevin claim to the kingdom of Naples by leading the First French Invasion of Italy and ousting Ferrante of Aragon from the Neapolitan throne. Ferdinand of Aragon in turn intervened in Naples in support of his relative. The resumption of the competition between France and Aragon for control of the realm of Naples had serious implications for the papacy. Alexander at first supported Ferdinand, but in 1498 he recognized Charles's heir Louis XII as king of Naples. That did not prevent Ferdinand from making significant gains in the war for Naples. In 1503 Louis was forced to send troops to reinforce his forces in Naples. Alexander died in August as the French neared Rome. Louis ordered his force to halt near Rome in order to sway the conclave. It had been two centuries since an outsider army had tried to control a papal election.

Alexander had suddenly taken ill on August 11, a day after he attended a dinner, at which Cesare and several other persons also fell ill. The pope rallied and for a time seemed out of danger. Ferrara's ambassador described the cardinals playing cards at his bedside while the pope kibitzed, but he died a week later. Since his corpse was reported to have turned black immediately and Cesare was also ill, contemporary belief was that both had been poisoned. Modern experts argue that no known poison causes such a pattern of death and that Alexander's death was from natural causes, probably malaria. Regardless of the cause, the pope was dead, and a conclave had to follow. Cesare, whom everyone had expected to name the next pope, was too ill to take an active role before the cardinals entered conclave. He later said that he was prepared for every eventuality at his father's death except that he would be ill when it happened. Alexander had reduced the power of the Roman nobles and curtailed their ability to control the conclave, while the presence of the French army north of the city and a Spanish army at the Neapolitan frontier only forty miles to the south negated each other.

The cardinals, however, expected to come under military pressure not only from France and Aragon but also from Cesare, who still commanded an army, officially the pope's but his in fact. They agreed to hold the conclave in Castel Sant'Angelo to reduce the threat of armed interference. Remodeling the fortress's apartments for the conclave and negotiations with Cesare to clear his troops out of Rome delayed its start. A French agent informed the cardinals that his king would regard it as a major affront if the conclave were held without d'Amboise and the French cardinals. Della Rovere sent a courier from Genoa to demand that the cardinals wait until he arrived. The cardinals solved their problem by delaying Alexander's funeral and entering conclave on the mandated tenth day after the funeral, on September 21. D'Amboise's party reached Rome in time; among the eight cardinals who were absent, only two were French. The thirty-nine cardinals were the most since the system of conclave began.

Louis XII had been preparing to get d'Amboise elected pope for years. When word reached him of Alexander's death, he set his plans into motion.

He expected to get the vote of della Rovere because of the favor he had shown della Rovere while he was in France. Louis had made a secret treaty with Cesare by which Cesare agreed to throw the votes of the eleven Spanish cardinals, who were more loyal to the Borgias than to the Spanish monarchs, to d'Amboise for French help in keeping his lands. Louis also sent a letter to his ambassador in Rome to pass on to the cardinals, which exhorted them "to consult his pleasure and make the Cardinal of Rouen Pope." The Venetian ambassador added: "Every blandishment, promise, and inducement is employed, together with implied threats against those who may ignore the request."[6] D'Amboise left for Rome with plenty of gold and orders for the French army to stay near Rome. He had with him Cardinal Ascanio Sforza, whom the French had taken prisoner during their occupation of Milan in 1500. Louis freed him in exchange for his pledge to vote for d'Amboise. When d'Amboise entered Rome late in the evening of September 10, a huge crowd was in the streets shouting his name and saluting him as the next pope. The Roman bookmakers, who accepted bets on who would be elected, did not think so highly of d'Amboise's chances. They set his odds at 13 to 100, while della Rovere's were at 15 to 100; the favorite was Cardinal Piccolomini, who was at 30 to 100.[7]

The French quickly found that getting their man elected pope was more difficult than they had imagined. Della Rovere had his own ambitions of being elected. He told the Venetian ambassador that he was there to look after his own interests and not those of anyone else. He would vote for d'Amboise only if his was the final vote the Frenchman needed to be elected, which, he added, he did not expect to happen. He said that if he were not elected himself, he was interested in making a good Italian pope and not one of the "barbarians," by which Italians of the era meant any non-Italian. He raised the fear that a French pope would take the papacy back to Avignon. D'Amboise soon found that this was the attitude of nearly all twenty-two Italian cardinals. Before d'Amboise realized that della Rovere was not supporting him, he allowed the Italian to convince him to order the French army to move farther from Rome as a sign of good faith to the Italians, thereby reducing the leverage that its presence created. Sforza lived up to his pact with the French by voting for d'Amboise but refused to use his influence with other Italians to get them to vote for him. Worst, d'Amboise discovered that Cesare Borgia was too ill to control the Spanish cardinals, who instead heeded the orders of their king not to vote for a French candidate. The antagonism among the Italians toward the dead pope was so strong that it was impossible for a Spaniard to be elected.

The cardinals in conclave first drew up a capitulation, which unlike earlier ones had no mention of reducing their number to twenty-four. It did include a provision that those cardinals who had annual incomes of less than 6,000 ducats would receive 2,400 ducats a year from the pope to allow them to live as befitting a church prince. Again the sources disagree over the number of votes that the *papabili* received in the first scrutiny, but the one regarded as the most accurate has della Rovere receiving fifteen and d'Amboise, thirteen. D'Amboise knew that several cardinals had pledged to vote for him only in the first scrutiny. Seeing that he had no hope of being elected, he turned to Piccolomini, who had received four votes in the first scrutiny. In the second scrutiny he was elected by accession after the ballots had been counted. In his uncle's honor he became Pius III.

The new pope had been elected in part because he was about sixty-four years old and sickly, having missed the scrutiny where he was elected because of illness. He was still a deacon, so Della Rovere ordained him priest and bishop before his coronation. All the ceremonies and festivities wore him out, and he died on October 18, 1503. The few days that he had functioned as pope gave promise that he would be reformed-minded. He had pledged not to add his nephew to the College, and he stuck to it despite pressure from his relatives at his deathbed. Contemporaries and historians alike have lamented the opportunity lost by his quick death. Barely a month after they had left the conclave, the cardinals, now totaling thirty-eight, were back in their cells. None had gotten far enough away from Rome not to return in time. The betting in Rome was on della Rovere.

Della Rovere had used the month productively. He met with Cesare Borgia and the Spanish cardinals, none of whom had voted for him the previous month, and promised to maintain Cesare in his possessions and the command of the papal army. Cesare brought over all eleven Spanish cardinals. Della Rovere also gained the support of d'Amboise, who accepted the dashing of his hopes, and regarded della Rovere as the most friendly to France among the Italians. Only Ascanio Sforza and his allies held out, but by the time the first scrutiny was held, they too had come around by lavish promises of favors. More than the required majority went to della Rovere's cell the night before to tell him they would vote for him. The next morning the first ballot was unanimous except for his own vote. (Henceforth, "unanimous" will assume that the pope-elect did not vote for himself.) He was so sure of his election that he had his ring of St. Peter designed beforehand. He said that he would be Julius II in honor of Julius Caesar; at least that is what those cardinals who fell afoul of his authoritarian ways later claimed. Since the cardinals did not draw up a capitulation, Julius II was elected pope in the shortest conclave ever; the cardinals had been locked in for less than ten hours. The new pope was about fifty-eight years old; his year of birth has never been established. He still was a vigorous, energetic man. He was known as an excellent politician but also hot-tempered, stubborn, and strong-willed. He had the diplomatic skills to match his will to dominate, and had experience as commander of the papal forces for his uncle, Sixtus IV. He had fathered a daughter, for whom he provided a good marriage but kept in the background.

Louis XII, who had reacted to the election of Pius III with an outburst of bile against Cesare—"That son of a whore has prevented Rouen from becoming pope!"—was pleased with Julius's election.[8] Louis did not realize the new pope's determination to rid Italy of the "barbarians," and allowed himself to be used as an ally against Venice, which had seized several cities in the Papal States. Once they had been recovered, Julius turned his wrath against the French, and in 1510 the gullible Louis found himself the target of a "Holy League" made up of the other major states of Europe. He responded by playing the religion card: He called a council to meet at Pisa to depose Julius and elect another pope. In late 1511 five cardinals—two Frenchmen, two Spaniards, and an Italian—opened the council in their own names and in those of four more cardinals who were sympathetic to their cause. Sixteen French bishops and some theologians and abbots were also present. Julius II had vehemently objected to calling a council, but now he responded to the one at Pisa by calling his own for the Lateran palace. The first sessions of the Fifth Lateran Council held under Julius were largely concerned with hurling anathema at the prelates in Pisa, Louis XII, and his allies in Italy. The pope deposed the cardinals at Pisa and ordered them excluded from the next conclave.

At a time when the "Warrior Pope" held the throne of St. Peter and general councils were being used as blatantly political weapons, it was appropriate that war would decide the outcome of the competition between pope and French king. Louis collected a large army to send into Italy, with Gaston de Foix in command. His orders were to destroy the papal-Spanish army and then march to Rome, where he would depose Julius and see to the election of Spanish Cardinal Carvajal, the senior cardinal at Pisa, who revealed that he would be Martin VI. De Foix crushed the papal-Spanish army at Ravenna on Easter Sunday, 1512, but was killed late in the battle. The battered French army floundered without his leadership, and Julius, who was preparing to flee Rome, decided to stay. Upon receiving word of the French victory, the council of Pisa voted to depose Julius as "a disturber of the peace and an obstinate author of schism." With the French army soon in retreat, the bold words from the council could not be transformed into action, and it soon broke up. By late 1512 France itself was under attack by English, Spanish, and Swiss forces. As Julius, *il Papa terribile,* lay on his deathbed in February 1513, he had every reason to be pleased with the way he had trumped the king of the barbarians. Future popes, however, had reason to curse him, for he had made a pact with the devil: He invested Ferdinand of Aragon with the kingdom of Naples in exchange for aid against Louis. Spain dominated Italy for the next 200 years and interfered in papal politics more than France had been doing.

Julius II died on February 21, 1513, without reinstating the four surviving cardinals of the Pisa council. He had erected four cardinals soon after his election, including two nephews, but after that he allowed for attrition to reduce the College. Ironically, as his election was highly simonical, he issued a stringent bull against simony in papal elections. A man elected simonically was not a valid pope, and nonsimonical cardinals had the right to invoke a general council and call on the secular powers to oust him.

When the conclave began in the Vatican palace on March 5, twenty-five cardinals were present of a total of thirty-one, not counting the four deposed ones. There were nineteen Italians among them, as the College had returned to being more narrowly Italian. Among the others were an Englishman, the first to attend a conclave since 1370, and a Frenchman. French Queen Anne, in her role as duchess of Brittany, had forbidden the Cardinal of Nantes to attend the council of Pisa, because she disagreed with her husband's use of the council against Julius. That had made him an acceptable negotiator between Julius and Louis, and he was in Rome for the conclave. The deposed cardinals appealed to Emperor Maximilian I to enable them to attend, but he did not get involved, and the College refused to reinstate them.

The cardinals went into the conclave without any clear favorites. The ambassadors and the Roman bookmakers indicated that there were as many as twelve thought to have a chance of being elected. The one achievement that had come out of the Fifth Lateran Council to that point was a decree denouncing simony in the papal election. Although it had no long-term impact, it was observed in this conclave. That meant that the wealthiest cardinals could not use their riches to buy votes. There were, however, other means of using status to gain votes, such as Giovanni de'Medici's offer of his nephew's marriage to Francesco Soderini's niece to win his vote.

After the opening Mass and sermon, the cardinals drew up a capitulation.[9] The usual demands of keeping the College at twenty-four members and requiring approval of two-thirds for decisions affecting individual cardinals were included. Julius's arbitrary decision making had offended the cardinals, and they added clauses that restricted the pope's power to control the College. They also required the new pope to continue the Lateran Council and to get consent of two-thirds of them to close it. When the first ballot was taken, a Spaniard, Cardinal Serra, received thirteen votes, much to the surprise of most, but he did not gain more.[10] It appears that most of those who had voted for him did so not because they intended to make him pope but to hide their real choices until they gained a better sense of who the strong candidates were. Because Serra was liked, he received most of such throwaway votes. By evening, it was clear that Medici was emerging as the frontrunner, despite receiving only one vote in the morning. When the dinner plate of English Cardinal Bainbridge was removed from the Vatican, its bottom was scratched with the names of Medici and Julius's nephew Riario, indicating that he expected one of them to be elected.[11] At the scrutiny the next day Medici was elected.

Although Medici was wealthy enough to have bought the election, all were adamant that simony had not been involved. Medici's reputation as a consummate politician and his experience as leader of Florence won the tiara for him. Many cardinals believed that as the head of a major Italian state, he would balance the influence of France and Spain. Unlike several recent predecessors, he had no children. The new pope was only thirty-seven years old but had been a cardinal for twenty-four years. Several cardinals were loathe to vote for him because his age suggested that he would have far too long a reign, but his physical condition reduced that fear. He suffered from a painful fistula that required that he be carried around in a chair, which made him appear more feeble than he really was (he was pope for eight years). Showing no emotion as the vote count made him pope, he declared that he would be known as Leo X. Still only a deacon, he had to be ordained a priest and a bishop before being crowned as pope. His coronation was moved up by two days, because the new pope, who was notorious for his faith in astrology, was told that the moon was in the wrong house on the original date.[12]

Leo was the first Florentine to become pope. Almost everyone was pleased with his election, even Louis XII, who said that good things could be expected to come from a good man. The new pope moved slowly to undo the excommunications and interdicts imposed by Julius, because to move quickly would have suggested that they had not been justified. The days when a new pope blatantly undid the decisions of his predecessor were over. In October 1513 Louis issued an edict refuting the Council of Pisa and accepting the Lateran as valid, and Leo declared that Julius's excommunication of the Pisa council's supporters had not been directed at Louis.

A year later Louis was succeeded by Francis I, who immediately resurrected the French claim to Milan. In September 1515 he soundly defeated the Swiss, who had established a protectorate over Milan. Fearful that Francis would march on Rome, Leo rushed word to the king that he was ready to negotiate. They met at Bologna, and in 1516 they agreed on the Concordat of Bologna for the governance of the French church. It conceded to the French king the right to appoint all French bishops and archbishops, but with papal approval. Another clause had more direct bearing on papal elections. The pope had the right to fill any episcopal vacancies that resulted when French prelates died in Rome. The French kings found themselves facing a dilemma as a result of that clause, which reaffirmed an old papal right. If they required French cardinals to reside in Rome, then the right of filling their French sees would fall to the pope when they died. If the kings required that they stay in France, then it was improbable that they could make it to Rome in time for a conclave and affect the outcome of the election. The cardinals objected to being asked to stay away from Rome, because the full exercise of their offices, especially drawing their revenues, could occur only while there. On two occasions the French king recalled his cardinals from Rome as a sign of displeasure with papal policy only to have the pope die soon after, leaving the French with little say in choosing the next pope.

Leo faced a plot to assassinate him led by Cardinal Alfonso Petrucci, who turned against him when Leo removed his brother from power in Siena. Petrucci and two other cardinals, who thought Leo was ignoring the College, were arrested in May 1517 and tortured. Petrucci was executed, and the others, while given mercy, lost their right to vote in the next conclave. This incident, plus the flight from Rome of two cardinals who were Petrucci's friends, led the pope to plan a radical changeover of the College. He named thirty-one new cardinals, the largest single promotion until 1946, when Pius XII surpassed it by one. Three were Medici kin; seven came from major Roman families; three were generals of major religious orders; three were papal musicians or artists; and another four had been serving as bankers to the papacy. Only four were non-Italians. Among the new cardinals were Adrian of Utrecht, who had been tutor to Emperor Charles V, and Thomas Cajetan, who took an active part in Reformation controversies, only four months from bursting out in Germany. Leo's ignoring the capitulation, to which he had sworn, finally made the point that whoever was elected was going to ignore it. The capitulation did not disappear, but the cardinals no longer attempted to control their number or increase their authority through it. Arguably Leo took the final step in asserting papal supremacy over the College. As individuals, cardinals continued to act as key papal advisers, but the College as a whole lost its claim to be the Church's senate.

Leo X died unexpectedly on December 1, 1521. Twenty-seven days later, thirty-nine cardinals entered conclave; only three were not Italian—two Spaniards and one Swiss. Nine non-Italians and one Italian did not attend, despite the delay before the conclave began, which resulted from the capture of a cardinal on his way to Rome, who was held for ransom. It provided time for the agents of Charles V, Francis I, and Henry VIII to spread around the huge sums that the three monarchs sent to Rome. Henry was ready to spend 100,000 ducats to get Cardinal Wolsey, his chancellor, elected, but he also regarded Giulio de'Medici, Leo's cousin, as acceptable. The English king asked Charles to support Wolsey as part of the Imperial-English alliance and send an army to Rome to force the cardinals to elect him.[13] Despite the sums that the kings spent to influence it, the College was hard pressed to find money to fund the lengthy conclave that ensued and had to mortgage the papal tiara to raise the money.[14] Few Italians were ready to accept a non-Italian. That fact persuaded Charles V to support Medici's election rather than Wolsey's. Many cardinals, including some of Leo's, opposed electing Medici, arguing that it would introduce hereditary succession to the papacy. Francis I had proclaimed a year earlier that he had a million gold *écus* to get a pro-French pope. It is not clear that he actually sent that money to Rome. Francis showed better understanding of the situation at Rome by supporting the election of pro-French Italians, such as the three Venetians. He claimed to have the votes of twelve cardinals secured for his candidates.

Medici went into the conclave with the support of fifteen or sixteen cardinals, but had little chance of gaining more. Charles's ambassador in Rome commented that, being new to the post, he was not familiar with the intrigues of electing a pope and he found repugnant all the avarice and lies that it involved.[15] He stated that every cardinal wanted to become pope. The bookmakers accepted bets on as many as twenty cardinals, although Medici had the best odds at 25 to 100. Cardinal Farnese's odds were 20 to 100. One observer noted as the doors were locked behind the cardinals that they were as badly divided going into the conclave as they had been in the 200 years since a pope had last come from outside the conclave. After the cardinals agreed on a capitulation, an exercise in futility as always, the balloting began on December 30. Medici conceded that he would not be elected, so he campaigned for his friend Farnese, who received twelve votes, all from Leo X's creations. Since Leo's cardinals made up twenty-eight of the thirty-nine electors, most felt that Farnese was as good as elected. The Roman mob plundered his house, and his odds with the bookies jumped to 40 to 100. Yet the next day, he had only four votes, as the cardinals tried out other candidates.

After the second ballot, a cardinal petitioned to be released from the conclave for ill health, although it was said that actually he was upset by the behavior of his fellows. It was only after a physician swore that the smoke and foul air in the conclave cells, which were sixteen feet by ten feet in size, were a danger to his life that the cardinals agreed by a two-thirds vote to let him leave. It is a sign of how seriously the cardinals were taking the obligations of conclave, since in the past cardinals were readily excused. As vote after vote took place, one per day, and they dragged into 1522, the ambassadors and the Roman people became increasingly restless. Secrecy about the conclave's activities was unusually good for so long an election, but the ambassadors, especially the Italian ones, had ways of finding out what was going on. Farnese remained the strongest candidate, and at the eighth scrutiny his supporters made an all-out effort to get him elected. He received twelve votes, and Medici asked for his election by accession. Eight or nine cardinals agreed to accept him, and one cried, "Papam Habemus!" It was expected that most of the cardinals would rush to salute him as pope lest they be left out of the distribution of the new pope's favors. Two cardinals adamantly opposed to Farnese demanded a formal counting of the votes by accession, and it became clear that he had come up short of two-thirds.[16] This maneuver discredited Farnese, and he dropped out of the running. Medici turned to others in his party, but none had broad support. Wolsey received eight votes, but his age, about forty-eight years, was held against him, although the English ambassador tried to convince the voters that he was over fifty.

On January 9, before the eleventh scrutiny, Medici addressed the cardinals on the deadlock in the conclave. He stated that he now knew his candidates would not be elected, and those from the other factions were unacceptable. He suggested the cardinals look outside the conclave. He said that he was ready to vote for the absent Adrian of Utrecht, a man esteemed for his piety and, he carefully noted, aged sixty-three. Adrian was clearly pro-

Imperial, which gave him most of the votes Medici controlled. In the scrutiny that followed, he received fifteen votes. When Colonna, who had been leading the opposition to Medici's choices, also declared for him, the floodgates opened, and Adrian received thirteen more votes by accession to reach the two-thirds majority. Since he was not a voter, he did not need two-thirds plus one. It took a long time before the cardinal announcing the news to the crowd outside made himself understood as to who the new pope was. The Roman people were flabbergasted to hear that a foreigner who was then in Spain had been elected and were upset that he did not have a house to plunder. Nor were they pleased that a "barbarian" had become pope, who would probably move the papacy to Germany and hand Italy over to slavery. Someone put up a "For Rent" sign on the Vatican.

Three cardinals were dispatched to inform Adrian of his election, who was in Spain serving as Charles V's viceroy. On January 24 he heard of his election through a private letter, but the official delegation delayed its journey for a month, hoping that something might happen to undo it, and reached him only in early March to get his consent. Word spread that he had died, and Francis I began to prepare for a new conclave. In fact Adrian was healthy enough for his age, although the burden of being pope would soon wear him down. He told the cardinals that he would keep his baptismal name and become Adrian VI. (He used Adrian, not Hadrian, as he is often called.) They did not bring St. Peter's ring along with them, in the hope that this would force Adrian to go to Rome to get it. He made it clear that he would reside in Rome, but handing over his duties to a new viceroy and other delays kept him from getting there until August 28. Only then did he truly begin to govern the Church, since he had promised the cardinals that he would make no decisions until in Rome. When historians refer to his brief reign, they usually do not note that his effective reign was in fact barely a year in length.

The new pope was an unusual choice in more ways than being the first non-Italian in over 100 years and the last for 457 more. He was a scholastic theologian who had never been to Rome, when a contemporary prerequisite for being pope was active patronage of artists and humanists. The way in which Adrian most differed from his predecessors was that he said Mass every day while he was pope. It is debatable whether the two previous popes ever celebrated Mass at all.

Francis I was deeply shocked that Charles V's tutor and viceroy had become pope, and he threatened to create a schism, but Adrian worked hard to remain neutral in the ongoing war between the two monarchs. French intransigence eventually forced him to side with the emperor. As a theologian, Adrian understood the issues being raised by Martin Luther far better than Leo had and set about to reconcile the German monk to the Catholic Church. Adrian's election also shocked the humanists and artists, who quickly found that he was serious about reducing the expenses of the papal court. Adrian for his part was shocked by the cardinals' lifestyle when he reached Rome, and he began to reduce their incomes. The cardinals reacted with outrage, and several talked of schism. It probably was fortunate for Adrian that he died in September 1523, worn out by the burdens of his office. Had he reigned longer, he might have ended up with much the same place in history as Urban VI, who began the Great Schism. Rumors that the French had poisoned him seem unfounded. Charles's ambassador told how shabbily the cardinals treated the dying pope, demanding he tell them where the papal treasury was hidden.[17] The Romans demonstrated their resentment toward Adrian by erecting a statue of his physician with "Liberator of the fatherland" written on it.

Adrian VI's health had been failing for several months, and the politicking for the coming conclave had been going on as long. Having sent a powerful army into northern Italy in 1522, Francis I expected to be in a position to dominate the conclave and see to the election either of Jean de Lorraine or, more likely, a French partisan among the Italians, perhaps Cardinal Fieschi. The French defeat at La Bicocca eliminated his military threat, but he was still determined to control the outcome. When word of Adrian's death came, the three cardinals in France were ordered to rush to Rome. The French defeat in turn meant that Charles V was in a better position to control the conclave's outcome. He again supported Giulio de'Medici, who was an advocate for imperial policy. Henry VIII was determined to secure Cardinal Wolsey's election, but the odds against an absentee non-Italian again getting the tiara were very high. Henry sent two different letters to his ambassador to distribute to the cardinals. One supported Medici, which was to be given to him first. If the ambassador saw from Medici's response that he did not want to be elected, then he was to distribute the second letter in favor of Wolsey and not spare any favors, especially among the young cardinals, who were the most needy.[18] Farnese was the favorite among the bettors, with Medici a close second.

Before the conclave opened on October 1, reports circulated that Medici could count on sixteen or seventeen votes. Colonna was regarded as the next strongest candidate. Thirty-five cardinals entered the conclave; six were absent, including Adrian's only promotion, a fellow Dutchman. A thunderstorm raged outside as the cardinals drew lots for their cells. Medici drew the cell under *Christ giving the Keys,* which was seen as a good omen, as Julius II too had drawn it. The anti-Imperial/anti-Medici cardinals, about half of the College, succeeded in holding off the first scrutiny until October 6, when three French cardinals arrived. They rushed into the conclave in their riding clothes and boots. The first scrutiny gave Fieschi, the French candidate, eleven votes, while Carvajal, the Imperial stalking horse, had twelve. Both sides promoted different cardinals in hope of finding someone who had the broad support needed for election. In the second scrutiny Gianmaria del Monte came within a vote of being elected, counting the written ballots and those by accession, but Medici broke his word to support him if he came that close, and del Monte did not get the final vote. As the conclave dragged into its tenth day, the Romans outside the conclave began to demand a pope. A conclave official told the crowd that the cardinals were ready to give them one "who lives in England."[19] The Romans, shocked by the idea, shouted back: "Give us an Italian pope, even if he is a block of wood!" Wolsey received twenty-two votes in one scrutiny, which shows that many cardinals were not as hostile to electing a non-Italian as it has often been suggested, but he never went higher.

On October 13 the Imperial party began voting for Medici, while the other side put forward Farnese as its best hope. Conclave secrecy was nonexistent, as the ambassadors reported daily with detailed accounts of the balloting and life in the conclave. The requirement that the amount and quality of food be reduced as the conclave continued was also ignored. The Roman bankers took bets on when the election would be completed, with odds of 60 to 100 against it happening in October. November arrived, and the cardinals continued their fruitless balloting. Medici held his party in tight discipline, while the French faction began to crack as the conclave dragged on. Colonna had strong ties to Charles V, but he despised the Medici and held a block of four votes against Medici. But when the French proposed voting for Cardinal Orsini, it was too much for Colonna, who hated the Orsini more. He agreed to vote for Medici, who went over twenty votes. In no time he reached twenty-seven as others rushed to support him, and on November 18, he was elected, taking the name Clement VII. Those who took odds of 80 to 100 that the conclave would not end in November lost heavily.

Despite lasting a month longer than the previous conclave, which had elected an outsider as pope, that of 1523 ended up choosing the man who had been identified as the favorite going in. It has been rare for the favorite to win in a lengthy conclave, when neutral or unknown cardinals almost always have been selected. Clement demonstrated his political skill, already proven in ruling Florence, as he kept his party focused on getting him elected through the forty-eight days of conclave. Most expected that he would be a great pope. As a patron of humanism and art, he lived up to that expectation, but in other respects he was a failure, with severe consequences for the Catholic Church.

CHAPTER VI

Conclaves in the Reformation Era

C lement VII's election in 1523 can be called the last conclave of the Renaissance. The new pope was very much in the mold of Renaissance popes, with his patronage of art and humanism, concern for the well-being of his family and city-state, and having a son. He had little comprehension of the issues involved in the Protestant Reformation. His goal in dealing with the religious passions in Germany was to prevent any reduction in the power and prestige of the papacy, not to make the reforms that might have mollified many, but certainly not all, of those who were following Luther. Clement stubbornly resisted calling a council despite the enthusiastic support for one from Holy Roman Emperor Charles V, who had played a major role in securing his election.

Despite his reputation as a superb politician, Clement failed to prevent two events that badly damaged the papacy's prestige. The first was the sack of Rome by Charles V's troops in May 1527. After the victory of Charles's army over the French in 1525 at Pavia, Clement let Francis I persuade him to join an alliance against Charles. The emperor dispatched his army southward to scare the pope into abandoning that alliance. In 1527 his unpaid troops mutinied and seized Rome to extract their pay by plundering the city. Aided by the Colonna, who hoped to depose Clement and make Pompeo Colonna pope, the soldiers subjected Rome to ten months of destruction, rape, and looting. The pope and fourteen cardinals found refuge in the Castel Sant'Angelo, which held out against the mutineers. In December

Clement escaped to Orvieto. The Imperial forces left in February, and he returned to a ransacked city. One positive result was the reconciliation between Clement and Cardinal Colonna, who tearfully embraced the pope and begged forgiveness for subjecting his native city to such terror.

The second episode was the matter of the annulment of Henry VIII's marriage to Catherine of Aragon. Any pope would have been slow to reverse Julius II's decision allowing Henry to marry Catherine, his brother's widow; but given the long record of annulling royal marriages, Henry had good reason to expect that consent would be forthcoming when he submitted his request in March 1527. In short order the Sack of Rome overshadowed Henry's "great affair," and Clement dared not offend Charles by granting the annulment, which would have removed Charles's aunt as queen of England. The pope delayed and delayed, hoping the problem might resolve itself through the death of one of the principals. Finally an exasperated Henry broke the ties between the papacy and the Church of England, a process that was complete when Clement died in September 1534. England had taken an active part in recent conclaves, perhaps simply as a result of Henry's ambition to make Wolsey pope. Now English influence was largely removed from the political calculus that went into electing a pope.

Rome was quiet in the days after Clement's death, and the conclave began almost on schedule. Clement's long illness made it possible to prepare well in advance for the conclave. When the cardinal serving as legate in France asked Francis I whom he would support for pope, he replied, "The Cardinal of Lorraine will tell you, when in conclave, whom I wish to be elected pope."[1] Thirty-five cardinals were locked into the Vatican. It was a foregone conclusion before the conclave began that Alessandro Farnese would become pope, having been *papabile* since 1513. He was regarded as neutral in the never-ending conflict between emperor and French king and was acceptable to both rulers, who had about equal strength in the conclave. Clement often spoke of his desire to see Farnese succeed him and obliged his nephew Ippolito to swear to elect him. At the first scrutiny, Farnese received twenty-three votes, but the tradition that on the first ballot cardinals often voted for friends who had no hope of victory kept him from being elected immediately. When the French party declared in his support, the matter was decided. On the second day he was elected unanimously. The conclave lasted less than thirty-six hours. It was one of the few times from Julius II's election in 1503 to the present when a cardinal who entered the conclave as pope-apparent emerged as pope.

The new Pope Paul III was sixty-seven years old, and his age was a factor in some of the votes he received. He complained that Clement had cost him ten years as pope. Since he lived to 1549, had he been elected in 1523, he would have been the first to surpass St. Peter's years as pope. His family was not originally Roman, but he had been born in the city; and its people greeted him as the first Roman pope since Martin V. Father of four children, shameless nepotist, and avid patron of artists and writers, he was still a Renaissance pope in key ways. Nonetheless, Paul was the first Counter Reformation pope.[2] He and the next seven pontiffs should be called Counter Reformation popes rather than Catholic reformers, since almost every decision on reform made by the papacy was intended to counter Protestantism. Of course, much of what they did had nothing to do with reform, whether Catholic or Protestant.

Like Clement, Paul refused to convene a general council unless he could control it. When he was at last sure that he could, Paul agreed to summon a council in 1544. The place chosen was Trent, which was in lands ruled by Charles V yet was Italian-speaking and on the south slope of the Alps, thus satisfying both emperor and pope. When the Council of Trent opened in December 1545, only four archbishops and thirty-one bishops were present. Attendance during this first set of sessions that lasted until March 1547 never exceeded 100 bishops, mostly Italians. Francis I refused to send any bishops, on the grounds that the French Church needed no reform, and if it did, he and his bishops would do it without interference. In March 1547 the outbreak of war in Germany between Charles and the Lutherans forced the council's suspension.

Other ways in which Paul III earned his place as the first Counter Reformation pope were his approving of the Society of Jesus and naming reform-minded cardinals. The Jesuits, who have often been called "the cutting edge of the Counter Reformation," included a fourth vow of obedience to the pope as a requirement of their members. In respect to creating cardinals,

Paul soon after his election gave red hats to two grandsons, one fifteen years old and the other seventeen. The younger, Alessandro, lived to 1589 and participated in seven conclaves. During most of them he was *papabile*. Paul also named fifteen-year-old Giulio della Rovere in 1547, who two years later became probably the youngest papal elector ever. The scandal such appointments caused was largely balanced by the quality of the rest of his cardinals. Among those that he erected were Gasparo Contarini, GianPietro Carafa, Reginald Pole, and John Fisher, although Fisher, a prisoner of Henry VIII, was soon executed. For the pope, Fisher's execution was the last straw in Henry's moves against papal authority. Henry was excommunicated, and England placed under interdict.

Although Paul was eighty-two years old in 1549, he was still vigorous until word reached him that his grandson Ottavio, whom he had made duke of Parma, had allied with the pope's enemies. It broke his spirit and his health, and after a four-day illness he died on November 10. Paul had erected so many cardinals, seventy-one in all, that the College totaled fifty-four members at the end of his reign, one above the ancient number of cardinals. The next four popes came from among Paul's creations. For the first time in three centuries there were no Colonna or Orsini cardinals. Paul III had appointed some cardinals because they were known as supporters of reform, but he also kept a delicate balance between the Imperial and French factions in the College. Fifty-one cardinals attended the conclave, but two left because of ill health before the last two entered. The final forty-nine voters constituted twenty-three votes for a French candidate, twenty-two for the Imperial side, and four neutrals. Because it was so closely divided, the conclave lasted seventy-two days. The rules of conclave were blatantly disregarded, and one conclavist declared that the conclave was more open than closed. A letter sent to a conclavist said that Charles V "will know when they urinate in this Conclave."[3] Ambassador reports; letters between rulers, their agents, and cardinals within the conclave; and diaries of conclavists provide amazingly detailed information about the politicking and machinations within it.[4] It was probably the best-reported conclave ever.

The conclave began nineteen days after Paul's death, which allowed thirteen cardinals to join the twenty-nine already in Rome for the *prattiche*.

Only two cardinals of the large French contingent of fourteen were then in the city. Henry II, who had succeeded Francis I in 1547, had kept his cardinals at home, so that they would not die at Rome and have their French sees filled by the pope, as the Concordat of Bologna stipulated. When news came of Paul's death, Henry ordered the French cardinals to rush to Rome. Cardinal Ippolito d'Este, the leader of the French party and *papabile* himself, successfully delayed the first scrutiny by several days. One way he did so was by wrangling over the capitulation, which lasted three days before one similar to that of 1523 was agreed upon. All but two of the French cardinals reached Rome before the conclave ended.

Because of the number of cardinals, the Sistine Chapel and five other halls in the Vatican were used for the cells created by wooden partitions, while the balloting took place in the Pauline Chapel. Although the rules allowed only two conclavists per cardinal, most had three or more, and at one point, a count of those inside the enclosure revealed 400 including the cardinals, or an average of seven conclavists per cardinal. Attempts to reduce the number were not successful. A guard of 5,000 men had been assembled to protect the conclave and keep Rome quiet, but the city was so peaceful that half of the force was dismissed after two weeks.

There were two substantial issues that made it crucial to both Charles V and Henry II to get their man elected. One was the resumption of the Council of Trent. Charles was determined to get a pope who would resume the council quickly, have it meet again at Trent, and issue a genuine invitation to the German Lutherans to attend. His purpose was to find a compromise religious settlement that would end the sectarian strife in Germany. Henry II for his part opposed resuming the council, since the religious problems in Germany distracted the emperor from his war with France and reduced his ability to raise troops and money there. Henry was also leery of extensive reform of the Church coming from the council, which might reduce his control over the French Church. The second issue involved the Italian duchies of Parma and Piacenza, which Paul III had given to two grandsons. Charles regarded them as Imperial fiefs. Henry's daughter had married one of the pope's grandsons, and the king was planning to use Parma as a base to attack Imperial-held Milan. The new pope would decide whether to leave the duchies in Farnese hands or return them to Charles.

The absence of all but two of the French cardinals in the first days of the conclave meant the Imperial faction had the upper hand. It seemed probable they would conclude the election before the French arrived. Charles V told his ambassador in Rome that while his first choice was a Spaniard, the Cardinal of Toledo, he would accept Reginald Pole, Pio Capri, and Giovanni Morone. He excluded every French cardinal and the Italians in the French party as well as the neutrals Marcellino Cervini, Gianmaria del Monte, and GianPietro Carafa. The Imperial cardinals in the conclave looked on Pole as their primary candidate. Most of the other cardinals also respected him, since he had been in exile since 1532 for opposing Henry VIII's religious policy. Alessandro Farnese, who was expected to be pro-French, stepped forward as the leader of the Imperialists after he struck a deal with Charles by which his brothers would keep their duchies. He proposed that the voting be done in public, not by written ballot. He believed that when it became clear that Pole was close to being elected in a public vote, enough cardinals would want to be on the winning side to declare themselves for him. Arguing over that lasted two days, with those holding out for secret balloting winning out. Thus it was only on December 3 that the first scrutiny took place.

The written ballots were deposited in a chalice on a table before the altar in the Pauline Chapel. Three scrutators examined each ballot to insure it was valid, and then Innocenzo Cibo, the senior cardinal, read the names on the ballots. The practice of having several names on a ballot was still in use. It made no difference whether a name was first or last on the ballot; when a cardinal's name appeared on ballots of two-thirds of the cardinals in the conclave, he would become pope. Pole received twenty-one votes at the first scrutiny. Much to the annoyance of the Imperial leaders, he refused to solicit more votes, saying that he would leave everything to God. At the next scrutiny Pole reached twenty-four votes, three short of victory. As a concession to the French party, the leaders of the Imperialists had agreed not to permit voting by accession for the first three scrutinies, so Pole remained unelected. Cardinal Este let the French ambassador know of the conclave's critical situation, and he rushed to the conclave door to demand that the cardinals wait for the arrival of the French cardinals, who, he said, were already passing by Corsica. In fact he had no idea where they were, but he bluffed his way into having his story accepted by declaring that the French would not accept a pope elected without their participation. The threat of schism was verified by a spy at the French court, who reported Henry II's warning that if a pope not acceptable to him were elected, the pope would not see the color of French money.

The false report that the French were close only spurred the Imperialists into frenzied activity to get Pole elected before they arrived. They wanted him acclaimed without a scrutiny, but Pole refused to become pope by "entering through the window." His supporters then secured the agreement of three cardinals who had not been voting for him to accede to him at the next scrutiny. They were so confident Pole would be elected that they ordered papal vestments tailored for him. On December 5 he received twenty-three written ballots, and two more cardinals, neither among those who made the agreement the previous evening, acceded to him. The Imperial leaders looked with eager anticipation at the three cardinals who had promised their votes, but they remained silent. Nor did anyone else declare for Pole. It was soon revealed that the three cardinals had promised the French party that they would vote for Pole only if he had twenty-six votes, and he was one short. The Imperial leaders thought they had that vote from del Monte, but Pole's conclavist had failed to visit him in his cell the night before to confirm his vote, and out of pique del Monte refused to vote for Pole.

It was rare for a cardinal to come so close and fail to be acclaimed as pope. Pole, it turned out, had opponents, even if tradition dictated that they acclaim him at this point. For the French, he was not only Charles V's partisan, he was also a native of the realm that was their traditional enemy. For some of the Italians who refused to vote for him, he was too young (forty-five years old) and a non-Italian; others deemed him too ardent a reformer. Carafa rejected Pole because he believed he was guilty of heresy in the doctrine of justification, and he persuaded Cibo not to vote for Pole. Having failed to make Pole pope, the Imperialists tried to get Toledo elected, but he had several of the same defects that Pole had in the eyes of those who refused to vote for the Englishman, and lacked his reputation for integrity and piety.

On December 11, the first four French cardinals arrived, led by Charles de Guise, who brought with him a bank draft for 150,000 gold crowns, presumably for bribes, and Henry II's instructions on whom to elect. Henry would have liked to see a French cardinal chosen or one of the two Italians related to his queen, Catherine de'Medici. He recognized, however, that none of them had broad support, so he favored Ippolito d'Este, who would later claim that he had a letter from Henry pledging to see him become pope "by love or by force." Word had reached the French court on December 17 that Pole had been elected, and Henry's joy when it proved false was enormous. With the French now present, the conclave was deadlocked. Pole slipped back to twenty-three votes, where he stayed for most of the remaining scrutinies. For the French, the longer the deadlock, the better, since there were additional French cardinals headed to Rome. The last French cardinal to arrive was the Cardinal of Bourbon on January 14. The fact that Henry had ordered aged, infirm Bourbon to attend the conclave indicates the importance he placed on getting a pope of his choice elected. The rulers carried on extensive correspondence with their agents in Rome about who was acceptable and who was not. The fact that the agents could inform their masters on exactly what was going on in the conclave was one of the scandals of that election, as was the ease with which the royal commands reached their partisans inside the Vatican.

Ambassadors and agents were not the only ones who had full knowledge of the conclave's proceedings; equally well-informed were the bookmakers of Rome. This conclave provides the most extensive information on the practice of betting on who would become pope. The sources for this conclave supply the betting odds on the *papabili* and the sums of money that changed hands. The Venetian ambassador in Rome in particular kept close track of the wagering. He reported: "It is more than clear that the merchants are very well informed about the state of the poll, and that the cardinals' attendants in Conclave go partners with them in wagers, which thus causes many tens of thousands *scudi* to change hands."[5] By the end of the conclave one bookie was reported to have made 20,000 *scudi* on the wagering. Del Monte had entered the conclave as the betting favorite at 20 to 100, but three days later Pole's odds were at 25 to 100. On December 5 Pole was the wagering favorite at 95 to 100 but fell back to 40 to 100 after his failure to be elected in that day's scrutiny. When the first group of French cardinals arrived on December 11, he was still at 40 to 100, but a month later his odds were 16 to 100. As the conclave dragged on, there were bets on when and if it would elect a pope. On January 22, the odds against the conclave ending in February were 50 to 100, 20 to 100 for March, and 10 to 100 that it would ever elect a pope.

On December 28 the rules of conclave were enforced more rigidly. The food for those enclosed was reduced and the windows closed to make it more unbearable. Two cardinals had been forced to leave, although one of them would later return. The effort to force a decision by making the conclave as unpleasant as possible slackened, and soon the meals and wine provided those inside were better than they had been at the beginning. The improved conditions did not prevent more cardinals and attendants from leaving, however, and many of the original attendants were no longer in the conclave by January. On the other hand, a canvass of the persons in the Vatican revealed that secretaries for the major secular governments were present. The French ambassador boasted to Henry II how easily he could communicate with Guise, the head of the French faction.

Both factions proclaimed that they were willing to die in the conclave before voting for a candidate for the other side. It came true for a cardinal who died on January 31. Accusations of poisoning led to an autopsy, and the anatomist who conducted it declared that he found such clear evidence of poison that he could not be more certain than if he had administered the poison himself. The dead cardinal was part of the French faction, and the loss of his vote came just as the French were beginning to hope that their policy of refusing to consider anyone outside of their party was about to pay off. With the support of several neutrals and members of the Farnese contingent, Cardinal Salviati, the French queen's cousin, came within three votes of election in late January. The boomlet for him subsided, and the deadlock continued.

On February 6, instructions arrived from the French court authorizing the French faction to vote for a neutral candidate. Guise began to negotiate with Farnese over the neutrals. Charles V's most recent letter had again rejected Carafa and Cervini but had not mentioned del Monte, whom he had included earlier among the undesirables. Taking that as permission to vote for him, the Imperialists agreed with Guise to present him as their candidate. As a Tuscan, del Monte had the support of the Medici. On February 8 the sixty-first scrutiny took place, and he received forty-one votes out of the forty-five cardinals still in the conclave. Four Imperialists refused to vote for him. The French were immensely pleased with themselves. By holding tight party discipline and refusing to vote for anyone in the Imperial faction, they had succeeded in electing a pope whom they believed favorable to France, since del Monte had never voted for Pole. Word of the election was rushed to the French court via a courier called "the Cripple," who reached it in five-and-a-half days after the result was announced. Henry II remarked that it was a fine conclusion to the matter. The cardinals for their part were reported as leaving the Vatican with haggard countenances. The conclave of 1549–50 remains a prime example of political interference in a conclave, not because it was more blatant than in many other conclaves, but because it was so thoroughly reported.

The sixty-three-year-old pope took the name Julius III, because, it was said, he thought of himself as having the same nature as Julius II. That proved to be untrue, as he was a weak and vacillating man, but he did emulate his namesake by becoming embroiled in a bitter feud with the French. Within a month they were lamenting that they had been deceived over the choice of pope. Julius sided with Charles regarding the duchies of Parma and Piacenza, reconvened the Council of Trent, and insisted on using every papal prerogative in the Concordat of Bologna to the fullest. Late in 1550 Henry II wrote to Julius that he deeply regretted "having caused the election of a pope so incompetent, unworthy, and pernicious."[6] In August 1551 Henry discussed with his royal council the possibility of creating a patriarchate for the French Church, to free it from the papacy. While the threat of schism remained only rhetoric on Henry's part, he refused to send any French bishops to the second set of sessions of the Council of Trent, which met from 1551 to 1552. He also recalled all of his cardinals from Rome, so the pope could not fill any bishoprics were any French prelates to die there.

Another complaint that the French had against Julius III was his refusal to appoint more than one French cardinal, although many cardinals had died in

1550 (perhaps because the long conclave had sapped their health). Louis de Guise was named to fill the vacancy created by the death of Jean de Lorraine, his uncle. Julius scandalized Rome when he gave a red hat to seventeen-year-old Innocenzo del Monte. The pope had adopted him when Innocenzo was a small boy, because he admired his pluck when the boy was confronted in a street at Parma by an ape that had escaped its cage. All sorts of rumors explained the true paternity of the boy, including accusations of incest between the future pope and a niece, although Julius was more notorious for gluttony than lust. When the pope proposed to make his protégé a cardinal, both Carafa and Pole objected, but that did not deter Julius. The new cardinal behaved so badly that the pope was soon threatening to defrock him. Julius's appointment of Innocenzo to a bishopic vacated by a French prelate's death at Rome was one reason why Henry was eager to keep his cardinals in France. Julius named three nephews to the College, and in all he erected seventeen cardinals, all Italians but two. None were from Charles V's domains, despite his submission of four names for red hats. The pope wanted to promote Francis Borgia, the Spanish Jesuit and saint, but Borgia refused the office.

As a veteran of the interminable conclave that had elected him, Julius was determined to reform the system of election. He commissioned three cardinals to draw up a reform bull. His death in March 1555, supposedly from a starvation diet intended to cure his gout, prevented him from formally proclaiming it, although the cardinals had approved it in consistory. When the cardinals met in the Vatican, their first order of business was declaring that Julius's new rules were not binding on them. Charles V and Henry II had ordered their cardinals to prepare to go to Rome a year earlier, when Julius was thought to be dying; but when he did die, both were caught by surprise. There were nine French cardinals, but two were too old to travel. Farnese, the leader of the Italians allied with the French, was legate in France and did not reach Rome in time. Charles's son Philip, then consort to Queen Mary Tudor, urged Pole, who had returned to England as legate, to rush to Rome, but he believed that his work in England was more important. Morone, another candidate in the previous election, was serving as legate in Germany. When the conclave opened, thirty-seven cardinals of a total of fifty-seven were present; only seven were not Italian. Forty voted in the final scrutiny.

Carafa, dean of the College, rushed Julius's funeral, because there was no money for a long one, and began the conclave on the eleventh day after the pope's death. Rome's bookies gave the best odds to Cervini, followed by Este, who was again Henry II's choice. Charles V supported Pole and Morone, but he did not exclude Cervini as he had in 1549. Royal instructions meant little, however, because the cardinals moved quickly to Cervini's election. When Carafa, who had opposed Cervini earlier, agreed to accept him, only Este, who badly wanted to become pope, stood in the way. Without the French cardinals, who were only then boarding ship at Marseilles, he lacked the votes to stop the election. He did manage to require that there be a formal scrutiny after the cardinals proclaimed Cervini by adoration. Adoration was described later in the century as taking place when "various faction heads lead a cardinal into the chapel, during the day or the night, and when they achieve enough votes, they elect and revere him."[7] The written vote was unanimous, except for Cervini's own, which went to Carafa. This seems to have begun the practice that the pope-elect would vote for the dean of the College when it was clear he had the necessary votes.

The new pope declared that he would keep his name, becoming Marcellus II. He was the last pope to keep his given name and the last until John Paul I to use one not already well represented in papal history. The previous forty-two popes used twenty-one names; the next forty-two popes used only ten. Born in 1501 near Siena, Marcellus had been promoted to cardinal by Paul III. His enthronement proved too fatiguing for his fragile health. He died twenty-five days after his election, on May 1, 1555. His death occurred too soon for him to reveal whether he would have been a zealous reformer-pope with an understanding of Protestant positions, as his actions as cardinal suggested.

Only forty days after they had entered the previous conclave, forty-three cardinals reassembled in the Vatican; two more joined them by the final vote. Five French cardinals, who had arrived after Marcellus's election, had remained in Rome. Este was again Henry II's choice, followed by two French cardinals and then Carafa and Pole. Pole's efforts to maintain peace between England and France had softened French opposition to him. Three additional Imperial cardinals attended this conclave, but Charles V was distracted by the peace conference going on in Germany that would result in the Peace of Augsburg, and was not active in promoting a candidate for pope. Pole remained in England, and his absence hurt his chances as it had a month earlier. Carafa, who was convinced that Pole was guilty of heresy, used his absence to undercut his support. The principal Imperial candidates each had opponents in their own party who refused to vote for them. Charles's ambassador in Rome wrote to him that with the expected arrival of an Imperial cardinal the day after the conclave opened, there would be twenty-three members in their party; but, he added, "since they are clergy-men and cardinals, no one can be sure of anything."[8] After the election, Toledo severely faulted the ambassador for mismanaging the election for the Imperial party, while he blasted Toledo for not preventing its outcome.

In the first two scrutinies, first Pole and then Morone were two votes shy of election, but Carafa had persuaded the cardinals not to allow voting by accession until after the third scrutiny. The French party's effort to get Este elected fell well short, and as the leader of the French faction Farnese recognized that he could not get Henry II's first three choices elected. He swung his support to Carafa, the dean. He began shouting, "The Dean! The Dean!" Quickly the entire French party and the neutrals acclaimed him as pope. Eight members from the Imperial faction soon did also, so he was three votes shy of the required majority. The Imperial leaders, knowing of the bad blood between Carafa and Charles V, tried to restrain the cardinals who had not yet acclaimed him from doing so and for a day they kept him short of election. Recognizing, however, that if three among them defected to Carafa and made him pope, they would incur his already legendary wrath by withholding their support, they agreed to swing their votes to him.

On May 23, 1555, Carafa, who called himself Paul IV to honor the pope who had made him a cardinal, was unanimously elected. When he rose from his chair to mount the papal seat, he appeared to some cardinals to be much taller than Cardinal Carafa had been. The Imperialists consoled themselves with the thought that at age seventy-nine the new pope would not live very long. So convinced were the Imperial leaders that there would soon be another conclave that they wrote to Prince Philip, who was beginning to manage his father's affairs, on how he should handle the cardinals of their party who had defected to Carafa early on. They recommended that the prince write each a letter that showed no irritation toward them but make it clear that any future act of this sort would require a reconsideration of the favors he and his father had been showing them.[9]

Paul lived long enough to make an indelible mark on the history of the Church. He was an adamant supporter of Catholic orthodoxy and created the Roman Inquisition to ensure that no heretical or quasi-heretical views were to be found in the Papal States, which was the area of its official jurisdiction. He prevailed on most of the Italian states to use it for their cases as well. His Index of Forbidden Books had a broader range of competency, acting to ban suspect books across Catholic Europe. Contrary to what has usually been said, he was not hostile to the idea of convoking a council, but he was adamant that it meet in Rome and not be a continuation of Trent. He believed Trent had approved of dangerous positions on key doctrines. One cardinal he regarded as responsible for those suspect decrees was Morone, who had served as legate to the council. In 1557 the pope ordered him arrested for heresy, and he was locked up in Castel Sant'Angelo for the rest of Paul's reign, refusing to admit any error or beg the pope for mercy. Paul had long accused Pole of heresy and in April 1557 demanded that Pole return to Rome for trial. He refused, but the accusation was a factor, along with the deteriorating situation for Mary Tudor's government, in Pole's declining health. He died the same day as Mary did in November 1558. With her death the break with Rome was again in effect, as Elizabeth I took the English throne.

Paul IV was determined to reform the papal curia and the College. He issued a draconian decree against simony and vigorously enforced Paul III's bull of 1547 allowing cardinals and other prelates to hold only one bishopric at a time. At that point Ippolito d'Este was archbishop of Milan and Ferrara in Italy and held Lyon and four more French sees. In general Paul was keen on reducing the princely lifestyle of the cardinals. He lived a blameless private life and rigidly enforced clerical celibacy. Even the laity in the choir of the Sistine Chapel were forced out, because they infringed on the realm of the clergy. One vice of his predecessors in which Paul did indulge was nepotism. He quickly gave a red hat to his nephew Carlo Carafa, a soldier, and gave another nephew the duchy of Paliano, taken from the Colonna. These unscrupulous, ambitious men directed the government of the Papal States and policy toward other states. Paul added two more relatives to the College; one was his eighteen-year-old great nephew Alfonso, who had a reputation for piety. Most of his eighteen cardinals were Italian theologians, canon lawyers, and superiors of religious orders. Only three were non-Italian, one each from France, Germany, and Spain.

The fact that even one of Paul's appointments was a Spaniard was a surprise. The pope hated Charles V, whom he blamed for the Sack of Rome, and the Spanish in general for their harsh rule in his native Naples. He also sharply criticized the emperor for making the Peace of Augsburg, with the Lutherans in which Lutheranism received equal status with Catholicism in Germany. Paul promised to give Milan and Naples to two of Henry II's sons in exchange for a French army to drive the Habsburgs out of Italy. Even before a French army arrived in Italy, war had erupted between Spain and the papacy in August 1556. A year later the Spanish army destroyed a French force at Saint-Quentin north of Paris, and Henry was forced to recall his troops from Italy to defend Paris. Within a month the Spanish were at Rome's walls. Paul, despite his rage, had to make peace with Philip II, who was now king of Spain after his father retired. Charles's brother Ferdinand became Holy Roman Emperor, thereby separating the Habsburg dynasty into the Austrian and the Spanish branches. For the next 100 years Spain dominated Italy and forced the papacy into a close alliance. Although the Spanish monarchs and the popes would have differences of opinion, they were largely over the details of how to implement the Counter Reformation, not its substance. Only a few popes in that time span were openly Spanish creatures, but there was no question that Spanish interests were paramount in the conclaves and the curia.

For the first three years of his reign, Paul IV astonished everyone with his energy and activity. In early 1559, however, he finally awakened to the duplicity of his nephews. He stripped them of their offices, but their betrayal sapped his health. He died on August 18, 1559 after a brief illness. At his election, the Romans had acclaimed him as one of their own, and his popularity increased when he spent lavishly for his enthronement. For all of his personal austerity, Paul spared no money for adorning the papacy. Soon his campaign to reform the morals of the Romans by banning gambling, prostitution, and blasphemy cost him his popularity. By the time he died, he was probably the most hated pope ever, not just in Rome but across the entire Catholic world. When word got out that he was dying, the Romans wrecked his house and those of his relatives, demolished the building of the Inquisition, burned its records and freed the prisoners, and destroyed his statues and coats of arms.

Before the conclave, the College decided to admit three cardinals whom Paul had intended to exclude. Carafa was permitted to return from exile and participate. Morone, who was still in prison, was released, and his right to vote restored. Paul had banished Este for simony and barred him from voting, but the cardinals welcomed him back. By these decisions the cardinals made it clear that a pope's authority to prevent a duly appointed cardinal from exercising the privileges of the office ended with his death. Forty cardinals entered conclave on September 5, a week late. Fifty-one participated in all, but because of illness and deaths only forty-eight voted at one time. Seven cardinals did not attend. It was the last conclave until 1958 in which non-Italians constituted more than one-third of the voters.

Thirteen of Paul IV's cardinals participated, led by his nephew. This was the first conclave in which the cardinal-nephew functioned as the leader of his uncle's cardinals, but they had no clear favorite and were open to persuasion to vote for a candidate from the other factions. Carafa had been in touch with Alessandro Farnese months before his uncle's death to discuss the upcoming conclave. Recognizing that neither would be elected, they agreed to work together to elect a pope beholden to both, thereby ensuring their fortunes under the new pope. As the balloting unfolded, most cardinals were surprised to find Carafa and Farnese working together. It was one manifestation of the cynical politics that often pervaded this era's conclaves.

There were sixteen votes in the French faction, including six French cardinals. For all of the French crown's services to Paul, he had named only two French cardinals. Henry II's death in a jousting accident the previous July had placed the crown on the head of his fifteen-year-old son, Francis II, and the Guise ran the government for him. Thus Charles de Guise, a *papabile*, did not attend, and French efforts to manage the conclave were less effective because of the unsettled situation at home. Este was again the French candidate, and he and Louis de Guise were the French leaders. The Spanish party consisted of seventeen cardinals, whose leader was Ascanio Sforza, and their candidate, Pio Carpi. Four cardinals were regarded as staunchly neutral. With the French party not yet at full strength on September 5, the Spanish hoped to get Carpi elected by adoration the first evening, without holding a formal ballot. The scheme was ruined by those cardinals in the Spanish party with their own ambitions of getting elected, who were not willing to have the conclave end without testing their chances. The cardinals then turned their attention to drawing up a capitulation. Its most interesting clause prohibited the pope from waging war on a Catholic prince, a clear reference to Paul IV's war with Spain. It also called for reconvening the council.

The scrutinies began on September 9, but no one rose above twenty votes for the first five days. Barthelemeo de La Cueva, a well-liked fellow but not a serious candidate, approached the others one by one, asking each to honor him with his vote. He and his conclavist had done this so carefully that thirty-two cardinals, a sufficient number to win, had agreed to vote for him, none having any idea that many others were planning to do the same. For several cardinals the joke was too good to keep to themselves, and moments before the scrutiny, they began to tell others, only to discover to their horror that most cardinals were planning to vote for La Cueva. His near-miss was treated with good humor. A week later a similar effort to honor Francesco Pisani, the dean, came close to getting him elected.

On September 13 Louis de Guise arrived from France with promises of money and property for Carafa if he supported a French candidate, although they were not enough to satisfy him. Now the politicking began in earnest. The French party continued to pick up strength with two more arrivals the next week. François de Tournon was the most respected French cardinal, and a concerted effort to make him pope secured him promises of more than enough votes for the scrutiny of September 22. Carafa got wind of the plan and derailed it by announcing that he intended to support Tournon, thereby costing the Frenchman the votes of several who detested Carafa and would not give him any role as popemaker. With the boomlet for Tournon over, the leaders of the French party turned to Enrico Gonzaga, a member of the Spanish faction but someone the French government found acceptable. Carafa and Farnese were adamantly opposed to Gonzaga, and their opposition was enough to keep him from the tiara. Francis II's letter favoring Gonzaga did not change the situation.

By now it was October 1 and the cardinals in conclave numbered forty-seven. Some cardinals left conclave and then returned, so the number of those voting fluctuated over the duration of the election. The rules of conclave had been well observed in its first weeks, but by late September they were largely ignored. An attempt to enforce them again lasted only a few days. The Roman bookmakers had no difficulty in keeping track of the balloting and adjusting the odds. It was reported that many conclavists and even some cardinals put down money on the election and kept the bookies informed. The worst violator of conclave was Spanish ambassador Vargas, who entered the Vatican as often as four times a day. His promises of money and positions to the cardinals in his king's name were so blatant that Philip II was accused of simony.[10] Vargas wrote to him on October 7 asking that he identify a new candidate, since Carpi's cause was hopeless. Philip knew of the support for Gonzaga, but his presence in the Spanish party did not ensure him of Philip's endorsement. He did not want to enhance the power of Mantua's ruling family, which could challenge his rule in Italy if the clout of the papacy was added to it. Thus he delayed a decision, and while he procrastinated, the cardinals dithered in conclave. The daily scrutinies continued, but many ballots had "no one."

A letter from Madrid on October 27 that said nothing about the candidates settled the election. It revealed that Philip intended to return the duchy of Paliano to the Colonna, from whom Paul IV had seized it for his nephew. Outraged by this news, Carafa swung his support to the French party. Philip's second letter a week later barred Vargas from working for Gonzaga. Vargas was admonished to let only Sforza, the leader of the Spanish party, know of this decision, but word got out quickly. Gonzaga withdrew his candidacy, but Philip's mistreatment of an admired prelate angered many Italians in the Spanish party. Accordingly Philip's order that his men in Rome continue to work to elect Carpi had little effect. The splintering of the Spanish party led Este to believe that the time was right to promote himself. But the Spanish faction was unified against him, and his efforts got him nowhere. In mid-December, with two Italian members of the French party dead—one dying in conclave and another outside—and a third out for ill health, Vargas and Sforza came within two votes of getting the Spaniard Pacheco elected. Their hopes that they had enough votes were dashed when again several who had promised to vote for Pacheco did not.

By now the cardinals had been in conclave for over three months. Vargas commented that they were so weary of the affair that they were ready to elect a block of wood as pope. The Venetian ambassador believed that the conclave would drag on for another six months. The factions had promoted all of their cardinals who were thought to have had a chance of being elected and had failed. Yet as had happened in other lengthy conclaves, the issue resolved itself very quickly once the cardinals stopped thinking strictly in terms of electing a partisan. Gian de'Medici, who was not related to the Florentine Medici but had close ties to Duke Cosimo de'Medici, had been voting with the Spanish party, but its leaders did not regard him as committed enough to support. He had gone into the conclave as one of the betting favorites, at odds of 16 to 100, following only Carpi and Este at 20 to 100. He routinely had three to seven votes in the scrutinies.[11]

From the first, Cosimo had promoted Medici and as the deadlock continued, exerted himself even more. Catherine de'Medici, queen mother in France, whose influence on Francis II was greater than generally thought, strongly recommended Medici. Philip II had not expressly excluded him, which in the convoluted world of the conclave was as good as endorsing him. Carpi, still hoping to become pope, refused to support him, and as long as Carafa remained uncommitted, Medici lacked a vote or two. The latter problem was solved when Medici made a solemn promise to support the Carafa brothers over Paliano, resist reprisals against them, and have the papacy pay their debts of 300,000 *scudi*. He was later accused of buying the election for that sum. After fevered negotiations on Christmas of 1559, enough cardinals agreed to accept Medici to acclaim him pope. In the morning a formal scrutiny verified the election. It was not unanimous, although the exact count has been lost. Carpi for one did not vote for him.

After a conclave of 103 days, there was a new pope. He took the name Pius IV, because he hoped to be what it signified. He was a sixty-one-year-old native of Milan. He rose in the Church as a canon lawyer. Paul III gave him a red hat in 1547. Pius had three illegitimate children before becoming a cardinal, but accusations that he fathered children as a cardinal are unproven. As pope he kept his children in the background, but that cannot be said about his nephews. He was one of ten siblings, and relatives by the dozens flocked to Rome to reap the rewards of being papal kin. A month after his election, Pius named two nephews and the teenaged son of Cosimo I as cardinals. One nephew was Carlo Borromeo, who was only twenty-one when he joined the College. From the first, Pius placed great faith in Borromeo's ability and loyalty, making him archbishop of Milan and governor of the Papal States. In 1562 another nephew received one of eighteen red hats given out. Two of Philip II's subjects and one Frenchman were the only non-Italians among them. A Colonna and an Orsini were promoted, ending a period of thirty years in which neither house had members in the College. At one point seven of his cardinals were under age thirty. Pius's youthful creations came under sharp criticism from reformers, who demanded a minimum age of thirty for a red hat. Borromeo proved worthy, but that was not true of his other young cardinals. Despite his conclave pledge, the pope had moved immediately to indict the Carafa cardinals and the duke of Paliano. In 1561, Carlo Carafa and the duke were executed, while Alfonso was imprisoned and died four years later. The harsh treatment of the Carafa was the only example of Pius's wrath.

Pius issued a constitution that concentrated on the conclavists. Each cardinal was allowed two, except for princes and the most senior cardinals, who were permitted three. Conclavists had to have been in a cardinal's service for a year before the conclave, but they could not be brothers or nephews. Collectively they were to be paid 10,000 crowns from the papal treasury and 2,000 more from vacant benefices. Pius forbade them from sacking the new pope's cell, but his decree failed to stop the practice. An "Advice for the Conclavist" from a century later describes the duties and rewards of being a conclavist. He was expected to take care of the physical needs of his cardinal, such as making his bed and keeping his food warm, but he was also to gather information and spread disinformation. He needed to have a hefty purse for distributing money to the minor conclave servants, who might be able to pick up key information. He was told to have wine and food available in his cell for entertaining. The conclavist also was advised to bring in disguises for himself and his cardinal, so they could slip unrecognized from cell to cell.[12]

The major event of Pius's reign was his reconvening the Council of Trent in January 1562. The final sessions were far better attended than the earlier ones, and even a French delegation arrived at Trent. The council defined doctrine, ritual, and governance for the Catholic Church that remained unchanged until Vatican II. One issue with which it did not deal was the reform of the conclave. Pius had given instructions to his legates presiding at the council to include it, but as was true of several controversial topics, it was passed over in the rush to conclude the council in late 1564 when the pope fell seriously ill. The cardinals were eager to end it before a papal election took place. They feared that if the pope died while it was still in session, there would be a powerful push to have the council elect the new pope. In December 1564 the council's work was quickly ended, and its decrees sent to Rome for papal approval.

Pius IV lived for another year, dying on December 9, 1565. Rome was quiet in the days after his death and remained so for the duration of the conclave. Some cardinals wanted to delay the voting to allow the French cardinals and Ugo Boncompagni, who was in Spain as legate, to participate. Borromeo successfully derailed that effort, because his uncle had issued a decree mandating the conclave start after ten days, and it began on schedule on December 19. Boncompagni, otherwise a strong candidate, remained in Spain. Forty-eight cardinals were then in Rome, and five more came before the conclave ended. Pius IV had named twenty-nine of the fifty-three present at the end, so Borromeo, custodian of his uncle's legacy, had unusual clout in the election. Morone, Este, and Farnese, *papabili* from previous conclaves, were the betting favorites as the doors locked behind the cardinals. The duke of Ferrara, Este's nephew, sent out letters proposing that it was time to elect a pope from a princely family, since the election of men of low status in the previous conclaves had created major problems.

Since the conclave began before the cardinals who were outside of Italy could reach Rome, there was only one non-Italian; eventually one Frenchman and three Spaniards arrived. The French and Spanish parties were made up almost entirely of Italians. Their leaders were Este again for the French and Pacheco for the Spanish. Philip II did not take an active part, unlike the prior election, telling his ambassador he only wanted a pope who was pious and peaceful. The religious wars were roiling France, and its monarchy could not exert much influence on the conclave. Catherine de'Medici, who wielded power in France, sought a pope who would compromise with the Protestants to help end the civil war. The emperor wanted a pope who was willing to allow clerical marriage as a concession to the Lutherans, which he had failed to get at Trent. Morone, with his reputation as both a conciliatory humanist and partisan of Spain, should have been the favorite, but by 1565 the appointments of Paul IV and Pius IV had largely transformed the College. Strict Catholic orthodoxy as defined by Trent was the mind-set of many cardinals, who regarded Morone as tainted by Paul IV's accusation of heresy. Since the rules of conclave were rigorously enforced, it was hard for ambassadors to make their masters' wishes known to the electors once the doors were locked.

As the most influential of Pius IV's three nephews in the conclave, Borromeo served as the leader of the large bloc of cardinals who owed their red hats to him. Borromeo had the responsibility of seeing that his uncle's wishes respecting the election were fulfilled as far as possible. Pius had issued a bull that insisted on strict conclave and barred factions in order to reduce political influence on an election. Borromeo, whom Pacheco had told that he would gain more merit with God by seeing to the election of a good pope than by a lifetime of fasting and prayer, tried at first to implement Pius's orders by doing little as leader of his party. Had he been more active, he probably could have secured Morone's election. In conclave's first evening, an effort led by the Spanish faction and approved by Borromeo to have Morone acclaimed as pope came close to success. If Morone had agreed to enter the Pauline chapel, it is likely that those cardinals hesitating in their support for him would have been caught up in the excitement of the moment and acclaimed him. Several cardinals who feared him because they had been involved in his citation before the Inquisition were white-faced with terror at the prospect. Like Pole in 1549, he refused to become pope without a formal scrutiny. That gave those vehemently opposed to Morone enough time to work on his tepid supporters before the ballot in the morning, when he was three votes shy of election.

Farnese now set about getting himself elected, but Borromeo regarded him as still unreformed and refused to support him. Demonstrations by the Romans outside of the Vatican failed to help his cause, and he did not go beyond sixteen votes. On December 31 Gonzaga fell mortally ill, and out of respect for him the cardinals stopped the balloting until he died. Borromeo then set about to elect Cardinal Sirleto, one of his uncle's red hats. His inexperience in the ways of the conclave led to the blunder of not informing the leaders of the other factions of his plans. Out of pique they refused to support Sirleto. Having learned his lesson, Borromeo met with Farnese, who proposed four new names. They agreed on Michele Ghislieri, who was rumored to be Philip II's choice, and Borromeo took his name to the other leaders. They acclaimed Ghislieri as pope on January 7, 1566, despite a report that four French cardinals were within a day of reaching the conclave. His acclamation as pope serves as a good example of the effect of peer pressure and perhaps even panic on those who might be opposed, lest one be on the wrong side of the new pontiff from the first. Even Morone, who in prison had faced Ghislieri as the grand inquisitor, acclaimed him as pope, making the vote unanimous.

Ghislieri announced that he would be called Pius V, to honor the uncle of the man he credited with making him pope, although Paul IV had made him a cardinal. He honored Paul by reversing the judgments against his nephews. Born in 1504, Pius came from a peasant family in Savoy. Joining the Dominicans he gained a reputation for zealously obeying the rules of his order. When he became pope, he kept the white Dominican robe, which was the origin of the white robe the pope wears today. He became attuned to Catholic orthodoxy when he joined the Roman Inquisition. Paul named him a cardinal in 1557 and then grand inquisitor. Only Adrian VI among the popes of the previous three centuries matched him in the sincerity of his piety. The cardinals complained that the Vatican and Rome were like a monastery during his reign, since Pius resumed Paul's efforts to make the city sin free.

Pius's frequent bouts of ill health kept rumors and intrigue active. He was well aware of the politicking among the cardinals for the next conclave and of the need to reform the electoral system. Instead of issuing a reform bull on the conclave, he decided to change the nature of the College by his appointments. The first one was his twenty-five-year-old nephew, Michele Bonelli, also a Dominican. The cardinals persuaded him to appoint Bonelli because the pope needed a trusted confident. Bonelli repaid his trust. Otherwise Pius kept his family at a distance. He took four years to fill the sixteen vacancies in the College, because he wanted to be sure of the merit of those he appointed. His creations included two French and two Spaniards, but the others were Italians, including Antonio Carafa, grandnephew to Paul IV. Most either had been at Trent or were leaders of their religious orders. Two came from great families, but even they were chosen first for their reputations as reformers.

As Dominican friar and inquisitor Pius V had little experience in worldly politics. Since he believed that God would protect the pope if that were His will, he greatly reduced expenditures for Rome's defense. He angered the Catholic rulers by blatantly interfering in their states and defended his actions with pronouncements that would have made Boniface VIII proud. Yet he had the greatest political feat for a sixteenth-century pope when he brought all the Catholic states except France into a league to resist the Turkish attack on Venetian-controlled Cyprus. Although the league failed to prevent Cyprus's fall, it won a great naval victory in 1571 over the Turks at the Battle of Lepanto.

Soon after word of the victory at Lepanto reached Pius, which he celebrated more lavishly than any other event of his reign, he fell ill. He lingered for four months before dying on May 1, 1572. Because of his effective campaign for reforming the Church and his devout life, he was canonized a saint in 1712, at a time when his rigid enforcement of orthodoxy was more appreciated than it would be today. He is the only pope between Celestine V and Pius X to be declared a saint. It has always been true that the nature of the papacy has required a politician more than a saint, and it is also true that the nature of the conclave has made it difficult to elect a saint. Certainly that was the case for the sixteenth-century conclaves.

Because of Pius's drawn-out death, the cardinals and Catholic rulers had time to prepare for the coming conclave; nonetheless, few non-Italian cardinals were in Rome. None of the six French cardinals attended, largely because of the chaos in France. Three other non-Italians and four Italians were also absent. Four Germans, two Spaniards, a Pole, and forty-four Italians made up the fifty-one voters. The last cardinal to arrive before the doors of the Vatican were locked late on May 12, 1572, was Cardinal Antoine de Granvelle, who while not a Spaniard was Philip II's choice as his party's leader in the conclave.

The election favorite was Alessandro Farnese, who had been campaigning for the tiara for two decades, although he always faced the formidable opposition of the Medici. He had led the French party in the previous two conclaves but since had courted Philip II's support. That enigmatic king did not indicate to Farnese how he stood before the conclave's start, but he communicated to Granvelle that Farnese was unacceptable. The story of how Farnese found that out is dramatic if probably apocryphal.[13] Rushing into the Vatican as the doors were being locked, Granvelle went straight to Farnese's cell and laid a letter bearing the Spanish royal seal on his table. When he reached for it, Granvelle yanked it back without letting him read it. He announced that Philip had excluded Farnese, but if he accepted gracefully, the king would support him in the next conclave. Out of consideration for Farnese's dignity, Granvelle said, he would rather not read the royal letter to the cardinals. If they could agree on someone to win on the first scrutiny, there was no need to embarrass Farnese in front of his colleagues. Farnese, at first stunned, recovered quickly and agreed, but only if he had a major role in choosing the pope. Granvelle suggested four names, and Farnese chose Ugo Boncompagni, who had impressed Philip as legate in Spain. Farnese then went to Bonelli and Borromeo, nephews of the previous popes, to get support for Boncompagni. Some in Borromeo's bloc preferred Cardinal Ricci, but Borromeo rejected him because he had a son. When after the election it was pointed out to him that Boncompagni also had a son, he said that he had not known. The cardinal querying him retorted that the Holy Spirit certainly knew and had not found it to be a barrier to making him pope.

In the evening of the conclave's second day, a formal ballot elected Boncompagni unanimously. He announced that he would be called Gregory

XIII in honor of Gregory the Great. Seventy years old, he was from a noble family of Bologna. He taught canon law for a time, and his students included Pole and Farnese. He went to Rome to serve Paul III and had major responsibilities for the next four popes, serving as legate at Trent. Shortly after its conclusion, Pius IV named him a cardinal. As his fathering a child indicated, he was not the austere monk that his predecessor was, and the cardinals and the Roman people hoped that his election would end the effort at reforming their morals. Gregory was an avid enough supporter of Trent not to allow much backsliding in church reform, but he was not the zealot that Pius V was. One issue that Trent mandated to the papacy was calendar reform, which was important to the Church because it involved the dating of Easter. Gregory accepted a new calendar that dropped ten days in October 1582 and made a small adjustment in the timing of leap years that corrected the problem. The Gregorian Calendar has given Gregory XIII immortality.

Gregory appointed two nephews as cardinals, but neither proved to be men of ability. As for his son, who badly wanted to become a cardinal or a prince, the pope gave him a position in the Papal States but never indulged his greater ambitions. Gregory wanted to reduce the size of the College, because it was too costly to provide so many cardinals with an appropriate income. He hinted that he would allow the College to drop back to twenty-four. In the first five years of his pontificate, he appointed only two more cardinals, both Austrian Habsburgs. In 1578 he named nine cardinals, only two of whom were Italians, in accordance with Trent's mandate that the nations be better represented. Four of the new cardinals were French, and three were from lands ruled by the Spanish crown. Gregory made no more appointments for another five years, and the College had dropped to forty-six when he named his next group. It proved to be a blockbuster event, as he gave red hats to nineteen men, including thirteen Italians. Four future popes were among them. Gregory's sincere effort to make the College representative of the Catholic nations proved futile. As European politics moved into what some tag the Age of the Baroque, the popes turned it largely into the reserve of Italian prelates. From 1585 to 1846, Italian cardinals routinely constituted 90 percent of the papal electors.

CHAPTER VII

Conclaves in the Baroque Era

By the time Gregory XIII died in 1585, the papacy had largely undergone the transformation called the Counter Reformation. He and the previous popes had dealt largely with shoring up the Catholic Church's position against the onslaught of Protestantism. While his successors hardly ignored the fact that much of Europe now professed rival versions of Christianity, they also turned their attention to internal changes in the administration of the Church in order to consolidate papal authority. Historians often refer to this era of increasing monarchial power as the Age of the Baroque, seeing in that artistic style a reflection of the interest in power that also appears in the political realm. The popes, beginning with Gregory's successor Sixtus V, largely created both the centralized bureaucracy for European rulers and, by their patronage of artists and architects, the Baroque style.

When Gregory died on April 10, 1585, there were thirty-eight cardinals in Rome, and four more arrived before the conclave ended. Those who participated included seven non-Italians. Eighteen cardinals, all but two non-Italians, were absent. Because of the many absentees, Gregory's efforts to broaden the College failed to produce a conclave with a stronger non-Italian voice, as Italians made up 83 percent of the voters, which was little different from previous ones. Alessandro Farnese, at his seventh conclave, was sure that it would be his last. Since he was now sixty-five and ill, he hoped that the cardinals would honor him with the papacy, while they could expect that he would have a short reign. He had not ceased campaigning since the previous election. He found out, however, that Cardinal Granvelle had deceived him in 1572 on both the strength of Philip II's objections to him and his gratitude for his stepping aside. Although Farnese's nephew was serving as Philip's governor in the Netherlands, Farnese garnered little support from him. In fact Philip was so slow in communicating his choices for pope to his ambassador that the conclave was over before his instructions arrived at Rome. Lacking instructions, his ambassador declared that the cardinals were to choose freely.

The factionalism based on the Spanish-French rivalry, which had dominated the conclaves since 1512, had largely dissipated by 1585. The religious wars had eliminated France as a player in conclave politics. When news of Gregory's death reached France, only one French cardinal began the journey to Rome, arriving too late to vote. Henry III's authority had come under severe attack by the Catholic League, and the one French cardinal already in Rome was a Leaguer who had no intention of taking orders from him. The French faction among the Italians had also dissipated. The role of balancing the Spanish party of thirteen cardinals fell to Cosimo II de'Medici, although he was an erstwhile Spanish ally. Cosimo's primary goal was preventing Farnese's election, but he also was eager to keep someone too beholden to Philip II from the tiara, and he had a skilled operative inside the conclave in his brother, a cardinal.

What factionalism did exist in the conclave was largely among the creatures of the previous three popes. The strength of Gregory XIII's group of cardinals was diluted by the absence of many of them and the meager political skills of Boncompagni, his nephew. Borromeo's death deprived Pius IV's cardinals of their astute leader, although Mark Sittich took his place. Pius V's promotions, led by Cardinal Bonelli, formed the most cohesive faction, but it also was the smallest. Although there was some differences among the three groups of cardinals, the main point of competition, in seeking to honor the man who had given them their exalted positions, was getting a protégé elected. The wagering odds among the Roman bookmakers put one from each party—Guglielmo Sirleto from Pius IV's cardinals, Felice Peretti from Pius V's, and Michele della Torre from Gregory's—as cofavorites at 12 to 100. Farnese was next at 5 to 100.

When the cardinals were formally placed in conclave the evening of Easter, April 21, 1585, they first drew up a capitulation that was very different from earlier ones. It required the new pope to continue the war against the Ottoman Turks, bring about peace among Catholic rulers, and see to the completion of the new St. Peter's, then seven decades under construction. There was little mention of the cardinals' privileges and none of reducing their number as in the past. At the first scrutiny Sittich, leading Pius IV's cardinals and supported by the Spanish, worked to have Sirleto elected, but many of the electors regarded him as too likely to become " the king of Spain's chaplain," and he failed to win. After the ballot was over, Cardinal Andrew of Austria appeared at the door. Some sought to keep him out on the basis that he had not yet been ordained a deacon, but Sittich got him admitted on the grounds that Gregory XIII had suspended Trent's rule that cardinals must be ordained in their proper office within six months of appointment. The additional vote for Sittich's candidate failed to make a difference.

The second scrutiny resulted in a broad scattering of votes. Seeing that the election was there for the taking, Medici began to canvass for Peretti. It was a surprise move, because Peretti blamed the Orsini, related by marriage to the Medici, for his nephew's murder. At this point Cardinal Madruzzo of Trent entered the conclave and assumed leadership of the Spanish party. He so actively supported Peretti that the Spanish faction assumed Peretti had received Philip's approval. Others found him acceptable because they believed him to be neutral. When Medici and Madruzzo counted noses and found that they had the two-thirds majority for him, they summoned Boncompagni and told him to require his uncle's cardinals to concur in order to have unanimity. He objected that Peretti was sixty-four years old and another election would probably be held soon. Frequent conclaves, he argued, did not benefit the Church. The others replied that Peretti was so vigorous that he hardly seemed to be fifty. With that Boncompagni yielded, and Peretti became pope unanimously except for his own vote for Farnese, the dean.

Felice Peretti chose the name Sixtus V to honor Sixtus IV, an earlier Franciscan pope. He came from a peasant family in the region of Ancona. His mother had a dream that her infant would wear the tiara, hence his name Felice. One task he had as a child was herding pigs, which explains the references to him as a swineherd. When someone used it to insult him, he retorted: "Yes, but if you had been a swineherd, you would still be one."[1] He joined the Franciscans, earned a degree in theology, and with his sermons in Rome caught the attention of Pius V, who made him a cardinal in 1570. He had a sour relationship with Gregory XIII, who did not believe in making members of religious orders cardinals. Gregory reduced his revenues and exiled him from Rome. Because he was absent from Rome, he was not well known to the cardinals. This enabled those who wanted to elect him to persuade others that his views were close to theirs.

Sixtus V surprised those who expected a reflective and ascetic pope. His first act was to attack the problem of banditry and lawlessness in Rome and the Papal States. The heads of hundreds of executed outlaws were staked on the bridge leading to the Castel Sant'Angelo. He angered the common people by greatly raising taxes. Higher taxes were hardly expected of a follower of St. Francis, but along with selling church offices, they restored solvency to the papal treasury. Sixtus was also an active patron of art and architecture in Rome. During his reign the dome of St. Peter's was finished, and he commissioned numerous churches and artworks associated with the Baroque style.

Sixtus's reign started badly, when a month after his election he named his fifteen-year-old nephew Alessandro as a cardinal. The boy was, to be sure, his only nephew eligible for the office, but he could hardly serve the pope as a trustworthy confident, which was the justification for promoting nephews. The purpose of such appointments was ensuring that the pope's family would have power and influence for a much longer time than the brief period that a pope could expect to reign. Several reform-minded cardinals protested, and one refused to attend young Alessandro's investiture, but with no results except to enrage the volatile pope. Sixtus soon demonstrated, however, that his young nephew's appointment was an aberration in a pattern of naming worthy cardinals.

The pope made his mark in conclave history by reorganizing the College and redefining the duties of its members. In December 1586, he issued a bull that fixed the size of the College until 1958. Because its size had been indeterminate since Julius II, having been as low as twenty-four and as high as seventy-six, Sixtus stated that he was setting the number at seventy in imitation of the seventy men who advised Moses (Numbers 11: 16–30). The increase reflected the need to have more representation from outside of Italy, but it also acknowledged the absence of many cardinals from Rome as nuncios. Seventy was not so much larger than the College's usual size under his predecessors that it reduced the cardinals' honors and revenues. The upper limit of seventy was deemed absolute; any promotion above that was invalid. But there was no obligation to keep the College at seventy; and during the next 360 years it often was less. Sixtus set the number of cardinal-bishops at six, one below the original seven; cardinal-priests at fifty, which required establishing an additional twenty-two Roman churches as their titular parishes; and cardinal-deacons at fourteen, four fewer than the traditional number. He made permanent the practice of giving the senior cardinal the bishopric of Ostia.

Sixtus repeated Tridentine mandates that cardinals be clerics and eligible to become bishops and have reached the age of thirty years, except for the cardinal-deacons, who needed only to be twenty-one but were obliged to be ordained within a year of their appointment. Bastard sons, even if legitimized, were barred from the College, as was anyone with children, legitimate or not. He extended the ban on the naming of brothers to the College to include cousins, nephews, and uncles of current cardinals. Since the College was required to advise the pope on matters of doctrine, Sixtus mandated that at least four cardinals have advanced degrees in theology. As the College also had the responsibility of representing the Catholic nations at the papal court, the decree repeated older mandates that it included men from all nations.

Sixtus also transformed the papal curia, creating the system of congregations that remained mostly unchanged until Vatican II. Some congregations already existed and were left untouched. Others, such as the Congregation for Rites, were new. There were fifteen in all, nine pertaining to the Church as a whole and six for governing Rome and the Papal States. Several cardinals were appointed to each, with one serving as president, except for the Inquisition, over which the pope himself presided. This arrangement systematized the handling of church business, which previously had been handled chaotically in the weekly consistory of pope and cardinals. The consistory was now held once or twice a month, at which new bishops or cardinals were announced. This change reduced the College's influence as a whole but enhanced the authority of those cardinals who presided over the powerful congregations. It also increased the power of the cardinal-nephew as administrator for the Papal States.

Sixtus appointed thirty-three cardinals, but only six were non-Italians, despite his intention of increasing their number. The most unexpected one was William Allen, the head of the English College in Flanders. He was named in August 1587 after Philip II asked for an English cardinal ready to serve as archbishop of Canterbury once his Armada had established a Catholic ruler in England. The Venetian ambassador reported the surprise in Rome at the appointment, especially among the bookies taking wagers on who would be named to the College. He wrote, "This unexpected promotion has caused incredible losses to the bookmakers of Rome."[2] The practice of betting on the making of cardinals is less well documented than for papal elections, but a document provides the odds for a consistory in December 1588. Odds of 90 to 100 were given on the naming of Cardinal de'Medici's secretary to replace him, since Medici had resigned to succeed his brother as duke of Tuscany.

In 1590 Sixtus indicated to his confidants his intention of reforming the conclave. He wanted to reduce the needed majority to half plus one and require that the final decision always be made by secret ballot, doing away with election by adoration and acclamation. He died on August 27, 1590, without mandating any change. Fifty-one of sixty-seven cardinals were in Rome when the conclave began; three more arrived before the final vote. Only six of them were non-Italians, including Cardinal Allen. When Sixtus died, the cardinals had to deal with the question of whether to summon three French cardinals to the conclave, realizing full well that unless voting went on for a month or more, they would never reach Rome in time. They had recognized the Protestant prince Henry of Navarre as king, who had succeeded to the French throne upon the assassination of Henry III in August 1589. The Spanish denounced them as schismatics, but most cardinals, noting that a cardinal's right to vote was virtually inviolable, agreed to send a fast courier to summon them. None tried to reach the conclave.

Most of the *papabili* from the previous conclave had died, notably Farnese in 1589 after fifty-five years as a cardinal. The creations of Gregory XIII and Sixtus V and the allies of Spain formed the three parties. The religious wars in France continued to keep the French out of the conclave. The twenty-two-man Spanish party included members of the other two divisions, and many of them had split loyalties; thus it was not a cohesive group. Gregory's fourteen cardinals were led by Francesco Sforza, who proposed that there was a tradition by which a cardinal appointed by the pope prior to the one just deceased was elected. His candidates were Giambattista Castagna, whom the bookmakers gave odds of 22 to 100, and Niccolo Sfondrato, 11 to 100. Sixtus's nephew, Cardinal Montalto, was only twenty-one years old at the time of the conclave, but he took to serving as the leader of his uncle's cardinals with zeal. Marcantonio Colonna, his candidate, was not his uncle's creation, but he was related by marriage. Again Florence had the most clout, as its new duke, the former cardinal and veteran of the 1585 conclave, was in an excellent position to exert influence.

There is no mention of a capitulation being drawn up after the cardinals entered conclave on September 9, for which they budgeted 250,000 *scudi*.[3] The first ballot the next morning revealed no trends. Cardinal Madruzzo again arrived a day late and assumed leadership of the Spanish party. He set to work to persuade Montalto that Colonna was unelectable, because he had children. On September 13, Colonna agreed to withdraw. Madruzzo then promoted two cardinals that Philip II had selected as his choices, but neither could get over half of the votes in the next scrutinies. The prospect of a deadlocked conclave led the faction leaders to caucus on who would be acceptable to all three parties. They agreed on Castagna, who had already received twenty votes on September 14. He had served as nuncio in France and Spain and was respected in both realms. Montalto agreed to accept him when the other leaders pledged to vote for one of his uncle's creations in the next conclave. Castagna's formal election on September 15 was unanimous. He announced that he would take the ancient papal name of Urban VII.

The new pope, who had received a red hat from Gregory XIII in 1583, was sixty-nine years old but still vigorous. He came from a Genoese family that had lived in Rome for a century, and he served as a conclavist in 1549 for his cardinal-uncle. The pleasure of the Romans at his election was enhanced by his decisions to reduce taxes and distribute to the poor much of the money hoarded by Sixtus. These were almost the only acts that Urban undertook, as he fell ill soon after his election and died on September 27. The day after his election he complained that the mosquitoes had been so bad he could not sleep, so he possibly died of malaria, common in Rome during late summer. In his will he left most of his wealth as dowries for poor girls. Only Adrian VI and Marcellus II among sixteenth-century popes compare in the amount of the charity mandated in their wills, and it is a strange coincidence that these popes had the three shortest reigns of the century. Urban's, at twelve days, remains the shortest since the eleventh century.

On October 6, 1590, the second conclave within a month began. It was so soon after the first that workmen had not finished returning the Vatican palace to normal. Except for Urban and a cardinal who had died, the same cardinals were locked in. The factions had not changed. Montalto was convinced that he would get Colonna elected this time, since he had Sforza's pledge of support for the election of one of Sixtus V's cardinals at the next conclave. Sforza, however, hated Colonna, and he honored his pledge by supporting another of Sixtus's men, Vincenzo Laureo. The Spanish ambassador had received Philip II's mandate for the previous election on the day it ended, and he acted on the assumption that nothing had changed in Madrid. Philip had named seven cardinals as his candidates and forbade his party from voting for thirty cardinals he deemed pro-French. Before, rulers had identified two or three cardinals whom they did not want elected, but none, not even Philip, had gone so far as to exclude a majority. The cardinals, indignant at this insult to their honor, complained that the next time, Philip might find only one cardinal acceptable and thereby dictate who would become pope. Montalto and Sforza vowed they would elect someone Philip had excluded, and a small pro-French party appeared. The Spanish party, however, consisted of twenty voters who remained loyal to Philip, and it proved impossible to elect a pope without them.

When the scrutinies began on October 7, Colonna was six votes short. Sixtus's rules made it impossible to steamroll the rest by accession, and in the privacy of their cells (where they marked their ballots in that era), enough voters remained convinced Colonna would make a poor pope to keep him from winning. Although rumors spread in Rome that Colonna was pope, and would-be officeholders rushed to his relatives to stake their claims, Sforza managed to prevent his election. Montalto then promoted Ippolito Aldobrandini, whom the Spanish leaders opposed because of his pro-French attitude. He came within three votes of the papacy but could not gain any more. As November arrived, the long conclave was leading to serious disorder in Rome, which was overwhelming 1,000 troops recruited to keep order.

The Spanish leaders felt the time was right to promote Gabriele Paleotto, who fell only three votes shy of victory. They pressed Montalto hard to swing the votes he controlled to Paleotto, but there was a different source of pressure on him. His mistress was Niccolo Sfondrato's niece, and she fancied having her uncle as pope. When Sforza proposed to Montalto that they support Sfondrato as the safest Spanish candidate, he immediately agreed. The two conspirators kept their agreement secret, and it was said that both made fortunes betting on Sfondrato, whose odds were 10 to 100 the day before he was elected on December 5. In memory of the pope who created him cardinal in 1583, he called himself Gregory XIV. A native of Milan, his widower father had become a cardinal in 1544. Gregory was fifty-five years old but was afflicted by gallstones, which were attributed to the fact that he drank water instead of wine. The new pope was devout, but he had no experience in administration. He named his nephew a cardinal and made him secretary of state. Unlike most cardinal-nephews, Paolo Sfondrato was pious, but he had no more political skills than his uncle. Their misadministration of the papal domains led to serious unrest. Both men served Spanish interests, and Philip received nearly everything he wanted from the papacy while they were in office, but the time they held power was brief. Gregory died in October 1591. He had issued a bull on the conclave that barred election by adoration and reinforced the rules of closure in order to reduce the influence of the great powers. Another clause banned wagering on papal elections, on promotions of cardinals, and on the length of papal reigns. This bull is the only source on the last sort of gambling. It is difficult to believe the bull had much effect on bookmakers and gamblers, but reports on the odds are less common after it, which has the result of reducing our sense of the ebb and flow of voting in the conclaves since then.

On October 27, 1591, fifty-six cardinals entered the conclave. Eight non-Italians were absent. Rome was already plagued by disorder, created largely by a grain shortage. For that reason alone the cardinals wanted a short conclave. Bitter complaints about Philip II's role in the past conclave had reached his ears while it was still going on, and he had sent a memo to his ambassador rescinding his exclusion of many cardinals, although he maintained it for the most pro-French ones. It had arrived after the election, but his ambassador took it as still valid. Montalto, still the leader of Sixtus V's cardinals, the largest party, had lost much of his clout because of his mishandling of the past elections. This conclave was perhaps the freest of the century.

The favorite going into the conclave was Giovanni Facchinetti. He had been among those Philip II had supported in 1590, but he was not perceived as ardently pro-Spanish. When Madruzzo, Philip's first choice, declared that he would not accept the papacy, Facchinetti had the upper hand. At the first scrutiny, he received twenty-three votes, and in the second, twenty-eight, half of the ballots. Montalto controlled enough votes to settle the issue. He had been supporting a cardinal whom he felt would be more generous to him, but he was easily persuaded that Facchinetti would be good to him. When he announced his support for Facchinetti during the third day, the election was over, although the vote was not unanimous. There is no explanation why the new pope chose the name Innocent XI. The former canon lawyer from Bologna was seventy-two years old and had resigned his bishopric twenty years earlier for health reasons. He was the third pope to come out of Sixtus's 1583 consistory. The belief that he would not reign long was a factor in many votes for him, but no one wanted to return to conclave after only two months, as happened when he died on December 30, 1591. The most pro-Spanish pope of the century, his short reign prevented him from doing much for Madrid.

For the fourth time in fifteen months the cardinals were in conclave. Four cardinals along with Innocent had died since the prior one, which suggests how hard even a short conclave was on the health of the mostly old men who participated. Fifty-two were present when the doors were locked on January 10, 1592. A cardinal soon died, but a Frenchman arrived two days later, so the voters remained at fifty-two. Because of the smaller number, neither of the two chapels in the Vatican was used for cells. Having had a taste of how useful a pope dedicated to Spain was, Philip was eager to win the election of someone similar and gave large pensions to many cardinals, which others denounced as simony. Philip's candidate was Giulio Santori, *papabile* in the previous three conclaves, who was a Neapolitan but pro-Spain. As the head of the Roman Inquisition he earned a reputation as a rigid enforcer of orthodoxy. Such a man had enemies, who deemed him another Paul IV, and the most outspoken was Mark Sittich, Pius V's nephew. No longer a callow youth, he would play a major role in this conclave.

Cardinal Madruzzo returned as leader of the pro-Spanish party. Sure that he had the votes to elect Santori pope by adoration, on the second day he led him to the Pauline chapel to become pope. Santori was so certain he was the new pope that he revealed he would take the name Clement, while the conclavists set about ransacking his cell. Sixteen cardinals refused to participate and remained in the Sistine chapel. Sittich was standing in that chapel's doorway as Santori's party passed on its way to the Pauline. Santori stepped forward to embrace him, but Sittich stopped him dead in his tracks, the story goes, by snarling at him, "Behold the devil's pope!"[4] Santori and Madruzzo were shaken by his vehemence, but they continued on to the Pauline chapel, where thirty-four cardinals were assembled, one over the two-thirds majority. Madruzzo told the dean that there were enough cardinals present to make Santori pope and wanted him so acclaimed. The dean insisted on counting the votes, and in the dim chapel lit by too few candles and with cardinals milling about, he could not get an accurate count. Ascanio Colonna suddenly shouted that he could never vote for Santori. As Colonna fled from the chapel, two cardinals tried to stop him and tore his robe in the process.

Down to thirty-five votes, Madruzzo still wanted to have Santori declared pope by adoration, but the dean, recognizing that a schism could follow from this situation, insisted that there had to be a ballot. Santori objected that he was already pope, but the dean replied that things had not reached that point. The sources disagree on whether the cardinals in the Sistine Chapel took part in the scrutiny, but when the ballots were counted, Santori had only thirty votes. Only two more cardinals gave him their votes by accession, leaving him three short of victory. He knew that his chance for becoming pope was gone. Brushing aside condolences from his friends, he rushed back to his ransacked cell bitterly disappointed, but within a day he recovered his equanimity and participated in the further balloting without rancor.

The Spanish party promoted more candidates, but Girolamo della Rovere, who was gaining votes as the conclave continued, died on January 26. It had only Ippolito Aldobrandini left. He was one of Sixtus V's cardinals, gaining him the votes of Montalto's faction. Sittich also found him acceptable, and on January 30 he was elected unanimously. As a Florentine, he became Clement VIII in honor of the Medici pope. He was fifty-five years old, a young pope by that era's standards. His deceased brother had also been a cardinal. Clement had to be ordained bishop before being consecrated pope. Despite coming out of the most rancorous conclave since 1549, the new pope worked well with the College's factions, since his role in the Spanish party had been marginal. The new pope was an opponent of nepotism, and he held off promoting any nephews for two years. But he came to feel that giving red hats to relatives was proper because a pope needed men whom he could trust absolutely, and he promoted his nephews Cinzio and Pietro. Pietro was twenty years old at his promotion, but he demonstrated real talent as a politician and became Clement's right-hand man. Cinzio's jealousy at this caused bad blood between the two nephews.

Clement VIII made the momentous decision in 1596 to accept French King Henry IV's conversion to Catholicism and absolved him from the excommunication imposed by Sixtus V. It is doubtful that any of his three predecessors would have done it, which restored a Catholic monarchy in France tied to Rome. Without Clement's open-mindedness on this matter, France might have gone in the direction of England, with an independent Gallican Church. The pope also annulled Henry's marriage to Marguerite of Valois and helped to arrange his marriage to Marie de'Medici, ensuring a Catholic succession in France. The vehement opposition of Spain to both steps suggests the pope's motive for accommodating Henry: He believed that the papacy needed a stable Catholic France to balance the power of Spain.

Spanish influence in the College in 1596 was greater than ever. The French cardinals had been reduced to two, while there were fourteen subjects of the Spanish crown from Spain and the Spanish lands in Italy. The two German cardinals were related to the Habsburgs, and five Italians were regarded as entirely committed to Spain. Clement set to work to undercut Spanish clout. He added five Frenchmen to the College, and the Italians he appointed were regarded as neutral. Clement pressed the French cardinals to reside in Rome, so they could take an active part in church governance and be there for the next conclave. Henry IV ordered three French cardinals to Rome, where they were when Clement died on March 3, 1605. The College was the largest yet for a conclave, at sixty-nine members, including thirteen non-Italians, but eight of them did not attend. Thirty-eight of Clement's creations took part in the conclave that began on March 14. There were nine from Sixtus's reign, whose leader was still Montalto. Pietro Aldobrandini assembled his uncle's creations on election eve and exhorted them to unity, an act the senior cardinals sharply criticized. Twenty-eight followed his lead in the first scrutinies. The Spanish party numbered thirteen, but Montalto was allied with them, and together they had enough votes to keep someone from being elected. The French party was eight strong.

In 1603, when Clement VIII had been seriously ill, Philip III, king of Spain since 1598, had sent a letter to his ambassador in Rome laying out strategy for the next *prattiche.* It says much about the sort of cardinal he supported that four of the six he named had died by 1605. While Philip II had the ability to express what was best for Spain in terms of what was best for the Church, his son spoke only of what was best for Spain. Italian anger at Spanish arrogance and interference vastly increased. Since most of the *papabili* from the previous conclave had died, twenty-one cardinals were given odds of winning by the bookmakers, who were back in business after apparently laying low since Gregory XIV's bull banning wagering on papal elections. The best odds, 10 to 100, were for Cesare Baronius, author of the first great Catholic history of the Church. He was Aldobrandini's first choice, whose uncle gave him the red hat.

When the first ballot took place on March 15, Baronius received twenty-three votes. It was a strong showing, since many of his supporters cast their votes for others out of friendship, as cardinals often did in the first scrutiny. Cardinal Avila, the Spanish leader, then revealed that his king had excluded

Baronius, to no one's surprise. Baronius had argued in his history that the supposed papal grants giving vast authority over the Sicilian Church to the king of Sicily, now the Spanish king, were forgeries. Avila raged at Aldobrandini, shouting that he would stay in conclave for a year to prevent Baronius from becoming pope. Aldobrandini shouted back that he would remain there for two years to get the pope he wanted.

A second altercation between the same two led to general shouting and shoving among the cardinals, which was taken by those outside as signs of election by acclamation, and they began opening the sealed doors and windows. After they were closed again, Baronius reached thirty-two votes, nine short of election, but he refused to canvass for more votes or to accept election by accession. Some cardinals in Clement's faction thought him too austere and refused to vote for him despite Aldobrandini's urging. By March 30 he despaired of making Baronius pope and began to promote others. Alessandro de'Medici quickly garnered broad support. The French strongly supported him, as did Duke Ferdinand of Tuscany, his relative. When Montalto swung his faction to him on April 1, his election was secured. Avila was caught by surprise by the turn of events and did not reveal his king's exclusion of Medici until many of his party had agreed to vote for him. At the final scrutiny, Avila shouted so loudly that his king would never accept Medici that he was heard outside, but few electors paid attention. Medici took the name Leo XI to honor Leo X, his great-uncle. He became the fourth pope from Sixtus's consistory of 1583. The French were pleased at his election, since he was kin to their queen; but their joy, and Spanish chagrin, did not last long. The seventy-year-old Leo caught a cold at a public ceremony and died on April 27, 1605. His only noteworthy act was a bull requiring secret balloting in the conclave.

The cardinals were back in conclave on May 8. Besides the pope, another cardinal had died in the previous month, leaving fifty-nine voters. An intriguing aspect of the study of conclaves is how quickly *papabili* changed from one to the next, even when the conclaves were only a few months apart. It was argued later in the seventeenth century that those cardinals who had worked against a failed *papabile* in the previous conclave would fear his revenge if he were now elected and would redoubled their efforts to prevent his election.[5] In 1605 the surprise was the support that Antonio Sauli, the choice of Montalto and Avila, had at the conclave's start. Henry IV deemed him acceptable; and many in Aldobrandini's faction were ready to vote for him, even Baronius, despite Sauli's refusal to support him a month previously. Baronius was a principled churchman rarely found in the conclaves, who was ready to vote for the man whom he thought would make the best pope regardless of faction or personal slight. Aldobrandini, however, rejected Sauli, who had opposed Clement VIII's election and had declared that the Church needed a pope willing to punish the cardinal-nephews for robbing the papacy. Aldobrandini controlled enough votes to ensure that Sauli would not reach the two-thirds majority. After the third scrutiny that was clear.

The next scrutinies saw Aldobrandini promote Roberto Bellarmine, who said that he would not lift a straw to get himself elected, and Domenico Tosco. Tosco had been a soldier and was still a man with many rough edges, albeit now a noted canon lawyer. His red hat was evidence of the range of men that Clement VIII had appointed. Avila also swung his party to Tosco, and early on May 16 his thirty-eight supporters led him into the Pauline Chapel to be acclaimed as pope. Baronius objected that Tosco was unsuitable and that his style of speech and dress would cause great scandal. Baronius's friends, emboldened by his courage, shouted that he should be acclaimed as pope. The two sides began to push and shove, and the tumult was heard outside of the conclave. Elderly Cardinal Visconti suffered several broken bones, the only known case of serious injury suffered in a conclave tumult.

After order was restored, the ensuing ballot showed that Tosco was two votes short of victory. No one gave him votes by accession, and his candidacy faded. Realizing how badly they had behaved, the faction leaders met to mend fences and restore dignity to the conclave. In discussing whom they could mutually support, Camillo Borghese's name came up. He was well liked but not regarded as papal caliber. The two leaders agreed to present his name to their factions. The French signed on, and the Spanish made no major objection. Late on May 16, 1605, the same day as the tumult, Borghese was unanimously elected. He chose to be called Paul V to honor Paul III, who had given his father, a native of Siena, a position in Rome. Camillo, a canon lawyer, was made a cardinal in 1596. He was fifty-two years old at his election and a robust man in good health. Many cardinals were not pleased that they had elected a young man, relatively speaking, for it reduced their chances of ever wearing the tiara. Unlike several earlier popes regarded as young, Paul did reign a long time, for almost sixteen years.

The cardinals soon had more reasons to be displeased with their choice. Paul ordered cardinals who were bishops of distant dioceses to leave Rome and reside in them. This reduced the influence and income of those cardinals, and the pope's rapacious relatives took over both. The most important of them was his sister's son, Scipione, who at age twenty-seven received the red hat and became known as Cardinal Borghese. He became the governor of the Papal States. By 1610 he had revenues of 153,000 *scudi,* whereas the entire family's income in 1592 had been a mere 4,900 *scudi.*[6] Other Borghese men received positions in Rome and the Papal States with huge incomes, while Borghese women were married into the best families, a prominent example of the marital policy carried on by most papal families of this era. Paul V was the most successful of any pope other than Paul III in establishing his family. While the Farnese pope had established his family as princes, its stay at the top was short compared to the duration of the Borghese atop Roman society to the late nineteenth century. The Borghese name remains splashed all over Rome, most prominently on the front of St. Peter's, although the Borghese pope had only a minor role in its completion.

Paul V had an exalted sense of papal authority over the secular rulers, which led to numerous confrontations with them. The most notorious involved Venice. Conflicts over control of church property and secular trials of clerics led Paul to lay an interdict on the city in 1606. Venice and most of its clergy ignored the interdict, and the pope, facing the possibility of a schism or even Venice becoming Protestant, was forced to accept a compromise. It was the last time that the interdict, the most powerful weapon in the papal arsenal, was used against a state. For Paul's contemporaries, the controversy with Venice was the major event of his reign, but what modern historians note more about his papacy is the Roman Inquisition's condemnation of Galileo Galilei in 1616 for teaching the theory of the sun-centered universe. It had been done without Paul's knowledge, but he acquiesced in the decision once he was informed.

Paul's long reign afforded many opportunities to create cardinals. Including his nephew, he gave the red hat to fifty-six men. Forty-two were Italians, and twelve were Romans. Several nominations recalled the worst days of the Renaissance papacy, such as that of the fourteen-year-old son of the duke of Savoy. But most blatant violation of the norms of the Council of Trent was the promotion at age ten of a son of Spain's Philip III. The fact that this cardinal-prince proved to be a churchman of merit hardly validated his being given a red hat before he reached puberty. Since Paul made his last creations a month before dying of a stroke on January 28, 1621, the College had seventy members, but twelve did not attend. Only three non-Italians were present. Several of Paul's creatures did not accept Cardinal Borghese's leadership, but he still controlled twenty-nine votes when the conclave began on February 8. Clement VIII's cardinals numbered thirteen, but his nephew Aldobrandini could count on only nine votes. Montalto, serving in his seventh conclave, controlled five of Sixtus V's. The national factions were still present, but with so few non-Italians taking part, Italian cardinals made up the French and Habsburg parties. They were poorly organized compared to the past; so while the antagonism between France and Spain was a factor in the conclave, the rivalries among the nephews of prior popes dominated.

Borghese was convinced that he had enough votes to break the tradition that the new pope should come from the cardinals of the dead pope's predecessor. Alessandro Ludovisi and Pietro Campori were most frequently mentioned as *papabili*, along with Sauli and Madruzzo, candidates in previous conclaves. Campori was Borghese's choice, and he was certain he had the votes to win Campori the tiara. The Habsburgs accepted Campori, and nearly all of Paul V's cardinals supported him. Campori, however, had foes: Venice ordered its two cardinals not to vote for any of Paul's cardinals; the French ambassador plotted effectively against him despite having no countrymen with whom to work; and Aldobrandini opposed all of Paul's creatures. Aldobrandini was so infirm that he was not expected to attend, but to Borghese's chagrin he was carried in on a litter.

Borghese believed that he could get Campori elected by adoration in the conclave's first evening. At that point there were fifty-six cardinals in the Vatican, and he felt certain of thirty-eight votes, enough by one to elect. The

French ambassador had a letter from Louis XIII excluding Campori, but he preferred not using it to avoid the rancor that an open exclusion always caused. Aware that two cardinals hostile to Campori were within hours of reaching Rome, the ambassador delayed leaving the Vatican despite all efforts to get him out. He left only when the two cardinals arrived after midnight. They voted as expected, and Borghese, realizing that Campori did not now have the required number, made no attempt to have him elected by adoration. When made aware of Louis XIII's strong opposition to Campori, other cardinals also withdrew their support. When it was time for the first scrutiny, Borghese decided not to promote Campori, hoping that a deadlocked conclave would turn to him later. Having put him aside, Borghese turned to Ludovisi, who was acceptable to France and Florence. He was elected by adoration, but whether by all is uncertain, for it is not known if the Venetians voted for him. The process of adoration did not leave a permanent record of the votes. The conclave took place in bitter cold; it is the first one for which cold was noted. A day after it ended, Aldobrandini and another cardinal died.

Ludovisi announced that he would be Gregory XV, to honor Gregory XIII, a fellow native of Bologna who had started him on his way up the hierarchy. He had just turned sixty-seven and was in poor health. In 1618 Paul V had named him a cardinal. Only three days after his election he made his twenty-five-year-old nephew Ludovico Ludovisi cardinal and secretary of state. Few complained about Ludovico's administration and the advice he gave his uncle, but the vast wealth he accumulated was scandalous. After only two years he purchased a great estate from the Colonna, who were slipping from their place in Roman society. After promoting only Italians up to 1621, the pope responded to the Catholic rulers' demands by naming a Spaniard and the Frenchman Armand de Richelieu. Gregory accommodated the Holy Roman emperor by promoting the Italian nuncio at his court, a way by which popes of this era satisfied both the monarchs and the desire to have a mostly Italian college. As his health failed in early 1623, it was expected that he would name more cardinals to enhance his nephew's power in the coming conclave, but no more red hats were forthcoming before he died in July 1623. It was reported to Richelieu that the dying pope told his nephew, who was pressing him to name more, that he had enough to account to God for the unworthy ones he had appointed.

Gregory XV's place in the history of papal elections rests on the bull that he issued in 1621. Most popes since Julius II had tried to reform the conclave, but they had been ignored in the conclaves, usually because the popes had died before putting the reforms in a bull. Gregory solicited a wide range of opinions. Borromeo and Bellarmine wanted to eliminate adoration and acclamation, because they violated secrecy and it was too difficult to verify the number of votes for a candidate. One radical plan proposed that the election begin two days after a pope's death and take place in the corpse's presence. It would have made only the top six vote recipients in the first scrutiny eligible for the second, and in each succeeding scrutiny without a decision, the weakest candidate would be dropped until only two were left. If the scrutiny with just two candidates failed to produce a pope, three cardinals chosen by lot would select the pope from the two men. While the plan had the advantage of ensuring the conclave would be over in a week at the most, it was too radical for Gregory, who took ideas on reform from his predecessors and made them comprehensive.

Gregory's bull strengthened the rules on sealing the conclave, making the French ploy of 1621 more difficult but not impossible. Once assembled, cardinals were obliged to swear that they would vote only for someone worthy of the papacy. The oath was intended to eliminate the frivolous votes for friends or as jokes. The bull eliminated election by adoration taking place without a secret vote, but it did not mention acclamation. The expected way for the election to take place was by written secret ballot, to reduce the pressure on cardinals to vote as demanded by a faction leader, although a contemporary commented that a cardinal's vote always became known eventually.[7] The ballot must contain the phrase "I choose as Supreme Pontiff my Lord Cardinal. . . ." Although Gregory did not ban voting for a non-cardinal, the ballot's format makes a powerful statement that such a vote would be highly irregular. To ensure its secrecy, it was to be folded in such a way that each of the three items to be written on it can be seen without the others being revealed. First, of course, was the name of the man for whom the ballot was being cast; only one name was permitted on a ballot, ending the practice of having several names on it. The cardinal, or his conclavist who filled it out, must use a disguised hand so that the scrutators could not discern whose it was. The voters must use separate desks, so others could not see the ballots. For most scrutinies the nominee's name was all that the scrutators needed to see. If, however, someone received exactly a two-thirds majority, then the clause that forbade a cardinal from voting for himself took effect, which ban was for the first time mandated. The voter must write his name on another part of the ballot and fold it out of sight so that it could be checked to see if the would-be pope had voted for himself. Should there be any dispute over that, the voter was also obliged to write on a third part of the paper a motto or a text that can serve as another check of whose ballot it was. Every word on the ballot must be in the same hand. Rules for getting ballots from cardinals too ill to come to the chapel were also laid out, to prevent any violation of secrecy. Any pacts or promises for votes were strictly forbidden.

Whether or not a cardinal could vote for himself had been interpreted differently over the centuries since it had first been banned in 1059. The winning majority had varied between two-thirds and two-thirds plus one, depending on whether one could vote for himself. The bull of 1621 firmly forbade it, which set victory at exactly two-thirds. When a draft of the bull was circulated among the cardinals, one objection involved Gregory's planned elimination of accession, which had shortened many deadlocked conclaves. He then allowed accession, but it no longer could be done verbally. After each scrutiny that failed to produce a pope, a cardinal willing to accede to a candidate, who must have received at least one vote, filled out a new secret ballot. Gregory sought to ensure that no one would put the same name on the ballot of accession as he had on the first ballot, thereby skewing the vote, by requiring the same motto on ballots of accession as used on the original ballot. Perhaps Gregory's greatest change was the move to two scrutinies a day. He intended not only to speed up the process but also to give faction leaders less time to negotiate and browbeat between scrutinies. They were allowed only to stand up immediately before a scrutiny and declare whom they wanted to promote. In March 1622 Gregory followed with a bull reducing the days for the dead pope's funeral to nine, so that conclaves would begin on the tenth day. The bull also required a mustering of everyone in the conclave except the cardinals at noon of the day following enclosure to ensure that everyone present belonged there. With these two bulls Gregory established the conclave format that remained largely unchanged to 1978.

If Gregory's intent was to streamline the conclave, the election that followed his death on July 8, 1623, demonstrated that he had not solved the problem. Thirty-four cardinals were in Rome that day, but more soon arrived, and the total by the end of the conclave was fifty-four. Four Spaniards and three Germans but no Frenchmen participated. The principal divisions among the cardinals involved the thirteen creations of the popes before Paul V, the thirty-two from Paul, and the nine from Gregory. The two popemakers were Borghese, who controlled twenty-seven votes, and Ludovisi, who increased his clout by allying with the cardinals from the Habsburg realms. Campori, Borghese's candidate in 1621, was again his choice. His age of sixty-six years was seen as an advantage, as revealed in a Spanish memorandum that suggested that an elderly man was more likely to be beholden to Spain and less likely to develop an independent policy.[8] Borghese expected that without much French influence in the conclave, the major obstacle to getting Campori elected had been removed.

If cold was a problem in the previous conclave, heat oppressed this one. Numerous cardinals took ill, including Borghese, whose confinement to bed reduced his control of the election. The conclave began a day late, on July 19, but the rules of enclosure were enforced better than in the past. After a first scrutiny in which the wide scattering of votes suggests that Gregory's ban on voting for friends had not taken effect, the evening ballot of July 20 revealed that Garcio Millini, not Campori, was the strongest of Borghese's candidates.[9] Ludovisi was adamant against Millini's election. He spread a story that Borghese had sworn that he would rather die in conclave than accept anyone from the other party as pope, which cost Borghese much good will, as the heat and filth in the Vatican wore badly on the cardinals as the conclave ran on.

As the favorites from both parties were put forward and failed, attention turned to neutral possibilities, among them one of Paul V's cardinals, Maffeo Barberini. Once he realized that he had a chance at the papacy, he actively solicited votes. Such personal campaigning was little seen in prior conclaves, as a cardinal in that situation allowed a friend or faction leader to do his canvassing for him. For a time his age of fifty-four years was held against him, but as others were promoted and failed, his chances rose. On August 4, Ludovisi agreed to accept him, which netted Borghese's violent opposition. The two men had become so estranged that they were not speaking to each other, but Borghese's illness, which had become so severe that he requested permission to leave the conclave, forced him to agree on a pope in order to end it. He accepted Barberini, who after all was his uncle's creature. It was arranged that he and his followers would vote for Barberini by accession on August 5. When the ballots were tallied, one was missing, although Barberini had enough for election. After three hours of debate over what to do, he asked for a new scrutiny. This time the tally came out right, and he received fifty of fifty-four votes. In the next two weeks eight cardinals died, as did some forty conclavists, while the new pope survived a bout with malaria contracted in the conclave.

Barberini, a Florentine, announced that he would take the name Urban VIII. Rumor spread that he had done this to show his affection for Rome, and the Romans greeted his election with great enthusiasm, as did the French court, where his election came as a pleasant surprise. Louis XIII could not have hoped that a pope so favorable to France would be elected. Barberini had served as nuncio to France under Paul V, and he owed his red hat to the French in 1606, after the Spanish had won one for the nuncio in Madrid. A canon lawyer, he came from a Florentine noble family. Its coat of arms included three bees, and Urban, whose family pride was enormous, spread the "Barberini bees" all around Rome, making it the most famous coat of arms of any pope. His sense of greatness led him to establish a vast building program in Rome, commissioning the Baroque architect Gian-lorenzo Bernini to design palaces emblazoned with the Barberini bees. So much of the building materials for his projects was stripped from structures surviving from antiquity that by the end of his reign there was a saying in Rome: "What the barbarians did not do, the Barberini have done." Urban set Bernini to work on such projects as St. Peter's Square, and they turned Rome into a Baroque city.

By the time of Urban's election in 1623, the system of papal elections had become nearly as regularized as the royal succession in the absolutist realms of Europe. Both Spain and France had only three kings in the seventeenth century, however, while there were eleven popes. Although the pope was not a hereditary monarch and no pope after 1605 was closely related to another, for the next 200 years they had highly similar characteristics. These popes were typically men of about seventy years who had spent ten years as cardinals after careers in church law. The odds were good that they came from noble families of Rome or the Papal States but not from the highest-ranking ones. Popes in that era, it can be argued, had more in common with one another than did successive kings in the hereditary monarchies—for example, Henry IV and Louis XIII of France; certainly the conclave never produced an incompetent like Charles II of Spain.

These popes paid lip service to Trent's goal of more non-Italian cardinals, but fulfilling it proved more difficult in this era than it might appear at first glance. The Dutch Republic, Scandinavia, England, Scotland, and much of Germany were Protestant, and there was little value in naming exiles as cardinals. William Allen's red hat in 1587 was an aberration resulting from the expected return of Catholicism to England under Spanish auspices. Bohemia, Hungary, and Catholic Germany were lands of the Austrian Habsburgs, and contemporaries believed the two branches of the family worked so closely that a cardinal from the Holy Roman Empire was in the Spanish king's pocket. From 1580 to 1640 the Spanish monarch ruled Portugal after the death of the last male in its royal family, and for three decades after it regained independence, no pope dared to offend Spain by giving a red hat to a Portuguese. When one was appointed in 1671, the Spanish cardinals boycotted the consistory. Switzerland was resolutely neutral, but with its cities Protestant, there were no prelates among the rural Catholics who might catch the pope's eye. In respect to Poland, it was so remote that most Italians wondered whether it deserved even one cardinal. France was the only option for internationalizing the College that avoided strengthening Spanish clout within it, and papal history well demonstrated the dangers of a powerful French party. In short it was not only Italian xenophobia and papal nepotism that led to so great a preponderance of Italian cardinals, but also the lack of an alternative to Spanish and French factionalism. Since much of Italy was under Spain, the search for a trustworthy College led to naming cardinals from the Papal States. After 1600 a fifth of them usually came from the pope's own realm.

This pattern of filling the College led to a division within it between the crown cardinals and the papal nephews on one hand and those who owed their red hats largely to long service as a curial bureaucrat on the other. The crown cardinals rarely came to Rome except for the conclaves, if then, and they were largely unknown to the majority of the College. Usually unable to take part in the *prattiche,* they were not *papabili* and rarely received more than one or two votes. The Italians and crown cardinals from other realms, of course, were determined not to allow a king to enhance his power by having one of his subjects become pope. With the cardinals equally adamant about preventing a papal nephew from becoming pope, the *papabili* for the next 200 years were almost entirely Italian curialists.

Chapter VIII

Conclaves in the Age of Absolutism

After 1600 most cardinals led respectable lives, and few had children. With men of good moral quality from whom to choose, the conclaves from then on elected popes who never disgraced their office with their personal failings. On the other hand, popes and cardinals were more eager than ever to enhance the status and wealth of their families by giving them much of the papal patrimony. Urban VIII set the example in this. An observer wrote, "Upon his elevation his kindred flew to Rome like so many bees to suck the honey of the Church."[1] His generosity and their rapaciousness made him the worst example of papal nepotism. His twenty-six-year-old nephew, Francesco, became a cardinal in October 1623 and later secretary of state. He became so important an adviser to his uncle that Urban called him *Padrone*, "Master." The cardinal-nephew became known as the *cardinale-padrone*. Urban's brother and another nephew, both Antonios, later received red hats. A third nephew received the highest positions in the Papal States open to a layman, and amassed so great a fortune that he rivaled the duke of Tuscany as the wealthiest man in Italy.

The cardinal-nephew was expected to lead his uncle's creations (whom he had a major role in naming) in electing his successor and into the future, but he was never regarded as *papabile* himself unless he outlived several popes. One reason was that he was usually too young and would be for several conclaves to come. Another was that he made many enemies, who preferred to blame the cardinal-nephew, not the pope, for adverse papal decisions. Perhaps most important, a pope such as Urban VIII would have seen to securing his family's fortune and permanent place in the Roman nobility. A cardinal saw no reason to enhance it further by giving the tiara to a second family member; it was the turn of another cardinal, perhaps himself, to elevate his family.

Upon his election in 1623, Urban reversed his two predecessors' pro-Habsburg policy in the Thirty Years War, which began in 1618, by sharply reducing the subsidy that the papacy had been giving Austria. Falling under the influence of Cardinal Richelieu, Urban refused to condemn France for continuing its long-standing anti-Habsburg policy, when it entered the war in 1635 on the Protestant side. A strong papal statement against the French decision might have prevented their direct involvement in the war. While Urban is sometimes cited as an enlightened pope because he denounced judicial astrology in 1629, the decision was not a sign of a loss of faith in astrology, but too strong a belief. He was outraged after astrologers predicting his imminent demise persuaded the Spanish king to send his cardinals to Rome for the coming conclave, and he banned predicting the death of popes.[2]

Urban's relationship with the cardinals was stormy. He ordered most to return to their dioceses, a move that cynics attributed to his desire to find more offices and revenues for his nephews. His relationship with Cardinal Ludovisi became so bitter that Ludovisi began to speak of holding a council to pass judgment on him. Ludovisi died in 1632 before anything came of it, and no one else had the standing to confront Urban's titanic temper. Urban's reign of twenty-one years gave him time to reshape the College. He promoted seventy-eight cardinals, but four were reserved *in petto;* their names were kept close to the pope's breast, and were not revealed before he died, which meant they never joined the College. Urban's cardinals included sixty-two Italians and twelve non-Italians; one was Jules Mazarin, the French minister but an Italian by birth. While many of the Italians were Barberini clients, a majority were men who had served as nuncios. Service in the papal diplomatic corps was becoming the surest route to a red hat. Although Urban largely ignored the cardinals, he allowed them to use the title of "Eminence" in order to outrank the Italian princes, who used "Excellency."

In June 1644, the seventy-six-year-old pontiff fell mortally ill, but Cardinal Francesco kept it secret from even his brothers for seven weeks. He tried to persuade Urban to fill the eight vacancies in the College to enhance his control of the coming conclave, but the pope refused to make promotions in such circumstances. When Urban died on July 29, fifty-six cardinals were in Rome; six missed the conclave that began on August 9. Among the fifty Italians, sixteen were Romans, and five were relatives of the dead pope. Only eight electors were not Urban's creations. Because of the deaths in the previous conclave, an effort was made to meet in the Quirinal Palace, which was newer than the Vatican and on high ground well away from the Tiber, but the Barberini quashed it. The cells in the Vatican were enlarged as a concession to the heat.

Never before was the bitterness between the French and the Habsburgs so great going into a conclave. The Thirty Years War was still raging, with France on the Protestant side. The French had actively incited revolt in several Spanish domains; Spanish rage at the French for stirring up rebellion was immeasurable. Spain was determined to get a pro-Spanish pope after the Francophile Urban. Antonio Barberini, the protector of French affairs, led in promoting Giulio Sacchetti, whose father had been a business partner with Urban's father. He was regarded as the strongest candidate. A Barberini outside of the conclave saluted Sacchetti's brother as "Your Eminence," indicating that he would become a cardinal. The brother called in friends to celebrate, but when it turned out that Giulio was not elected, he allegedly went mad.[3] Sacchetti's defects were that he was not yet sixty, which promised another long reign, and he was a friend of Mazarin, first minister in France since Richelieu's death in 1642. Mazarin for his part provided gold to ensure that Giovanni Pamfili not be elected because of his Spanish sympathies.

Francesco Barberini, the leader of his uncle's cardinals, supported Sacchetti and Pamfili, both Urban's creations. Cardinal Albornez, the leader of the Spanish party, announced before the first scrutiny that Philip IV was exercising his right to exclude Sacchetti from the papacy. When pressed to give a reason, Albornez would only say that his king did not trust him. The conclave was thrown into an uproar. The theologians in the conclave were consulted on whether the cardinals need pay attention to a king's demand. They agreed that since disregarding the will of a powerful king would imperil the Church, excluding a worthy cardinal was the lesser evil. Many cardinals refused to concede Philip that right, but enough agreed to the exclusion to make it a fait accompli. Sacchetti had only twelve votes in the first scrutiny.

Word on Sacchetti's exclusion was rushed to Mazarin, who rushed word back that he excluded Pamfili, but it came too late to affect the election. A deadly heat wave hit Rome in September, and malaria, the scourge of late Roman summers, struck many cardinals. One died on September 7, and others, including Francesco Barberini, left the conclave. Serious unrest also had broken out in Rome. The prospect of a drawn-out conclave under such conditions was terrifying, and Antonio, now leader of the Barberini cardinals, decided that he had to put aside partisanship and elect a pope as quickly as possible. That meant supporting Pamfili, whom Florence also backed.[4] Negotiations with the Spanish party secured an agreement by which the Barberini were assured of Spanish protection against reprisals by the new pope. On September 15, 1644, Pamfili was elected with five votes against him. Mazarin was so upset at the election that he threatened not to recognize Innocent X as pope. The pope gained some revenge when he gave a red hat to Jean de Retz, Mazarin's bitter foe. Mazarin threw Retz in prison, but he escaped and made his way to Rome, where Innocent protected him. The Retz affair caused bad blood between Rome and Paris throughout Innocent's reign.

The new Innocent X's family had arrived in Rome during Innocent VIII's reign. A seventy-year-old canon lawyer whose uncle had been a cardinal, he had received a red hat in 1627. He had fewer relatives than Urban, but he was just as eager to establish them in the Roman elite. He quickly gave a red hat to his only nephew, who reverted to lay status two years later in order to marry. It was Olimpia, the widow of the pope's brother, who wielded a cardinal-nephew's clout. Donna Olimpia, whom Roman punsters tagged *olim pia*, "formerly pious," was smart and ambitious. Innocent made no decisions without her advice, and everyone who sought an audience had to see her first with gifts in hand. Not since Marozia Theophylact had a woman exercised so much influence on the papacy. Olimpia secured red hats for two of her relatives, who disgraced their offices, but her influence was mostly beneficial to Innocent.

The major event of Innocent's reign was the Peace of Westphalia in 1648, ending the Thirty Years War. He had not been asked to send a representative to the negotiations or approve of the terms that gave official recognition to Protestantism in Germany. He demanded that the Catholic rulers renounce the treaty, but they ignored him. In response he gave out red hats to only six non-Italians out of forty appointments. Except for Retz, they were all crown cardinals, the term for those who received their offices at the insistence of Catholic secular rulers. One was Mazarin's brother, Michele. Despite Innocent's oath in conclave to protect the Barberini, he quickly seized properties and revenues from them. They successfully appealed to Mazarin for protection and became entrenched in the French faction, which reduced the impact of Innocent's pro-Spanish cardinals in the next conclave. The pope made peace with the Barberini by giving Carlo Barberini a red hat in 1553.

When Innocent died on January 7, 1655, his sister-in-law refused to pay for his burial, so it took eleven days before his body was interred in St. Peter's with no ceremony. The conclave began two days later. The death of a cardinal a month earlier left the College one short of seventy, and sixty-six participated in the conclave. Among the electors, Urban VIII's and Innocent's creatures both numbered thirty-two. The Spanish party had eighteen members, but the smaller French faction included Francesco Barberini, who was capable of bringing as many as twenty neutral cardinals to the support of the right candidate. The absence of a *cardinale-padrone* among Innocent's cardinals meant that they lacked the coherence of the other faction. Donna Olimpia was their real leader, but her exclusion from the conclave prevented her from controlling it. She, however, was invited into the enclosure to address the cardinals, the only woman so honored.

Giulio Sacchetti was again the strongest *papabile,* but the fact that the Spanish had excluded him as too pro-French in 1645 was held against him, although there was no new word from Madrid. In the early balloting, he regularly received thirty-three votes, giving him the nickname of *Trentatre.* More unusual was the number of votes for "no one," which reached twenty-seven on January 22. Most came from Innocent's cardinals, who did not want to vote for one of Urban's creatures but had no strong candidate themselves. Eleven of them made up what became known as the "Flying

Squadron," because they would fly to whichever faction promoted a candidate who stood for the best interests of the papacy and not one of the secular rulers.[5] Innocent X appointed them all, and they were both young and talented compared to the rest of the College; most came from the Papal States or Rome. Aware that none of them were *papabili* because they were too young, they were free to concentrate on electing their candidate, Fabio Chigi, the most respected of Innocent's cardinals, although many regarded him also as too young at age fifty-six. When a cardinal arrived three days late with a letter from Philip IV excluding Sacchetti, it sparked another heated debate over whether kings had a right to exclude anyone. Those who argued against it had numerous papal decrees to back their case, but those who supported the kings made the irrefutable argument that ignoring the will of powerful rulers risked schism. The candidates that Philip supported were so poorly regarded that they received nary a vote beyond the Spanish party.

Trentatre continued to get his thirty-three votes, as his supporters refused to concede to Spain, while the conclave dragged on into February. The younger cardinals became rowdy and played pranks on the older ones. An elderly cardinal died from pneumonia, reputedly after being left lying on a cold floor upon being frightened by a young cardinal dressed as a ghost. Others became ill in ordinary ways and left the Vatican. In mid-February Sacchetti, recognizing that he could not be elected, on his own initiative wrote to Mazarin and asked him to order the French fraction to support Chigi. Chigi had earned Mazarin's hatred when Mazarin had been exiled during the French uprising (1648–1651), known as the Fronde, to Cologne, where Chigi was nuncio. He had done little to help Mazarin, who never forgave nor forgot. After six weeks Mazarin's reply reached Rome. He begrudgingly agreed to allow his clients to vote for Chigi if it were impossible to get Sacchetti elected. He begged his supporters to stop voting for him and turn to Chigi, who had been routinely getting fifteen to eighteen votes. During the night of April 6, 1655, the Flying Squadron stayed up all night to ensure that no last minute hitch would impede Chigi's election. The next morning, he received twenty-written ballots, and the rest of the cardinals accepted him by accession. After eighty days the conclave was over.

Alexander VII, a name Chigi took to honor Alexander III, also from Siena, came from a banking family that had fallen on hard times. He earned a degree in theology and became a cardinal in 1652. His first move as pope was ordering Donna Olimpia out of Rome; his second was putting a coffin in his bedroom as a reminder of his mortality. He was convinced that nepotism harmed the papacy and was determined to avoid it. He said that as a cardinal he had many relatives, as pope he had none. The cardinals pressed him to bring his nephews to Rome not only to justify their own nepotism but also to provide the pope with his own confidants. For two years he held out; then relenting, he gave his relatives positions in the papal government and a nephew the red hat. By late in his reign his family had gained as much as any previous pope's relatives. Unlike most other papal families, which lost their lofty place in Roman society over time, the Chigi have remained prominent. In 1721, after the extinction of the House of Savelli, the Chigi received the honor of serving as conclave marshal, a post they have held to the present.

During Alexander's reign relations with France deteriorated further. A violent confrontation between papal troops and the French ambassador in Rome led to a French occupation of Avignon in 1664. They withdrew only after the pope abjectly apologized. Mazarin pressed the pope hard on creating French cardinals, but Alexander's nominations changed little from those of Innocent and Urban. Alexander promoted forty men, all but seven Italians. With six cardinals Siena profited from having a native son wear the tiara.

The College was at full strength when the pope died on May 22, 1667, after a long illness, which allowed time to prepare for the coming conclave. The first issue that the College faced was where to hold the election. The younger cardinals asked that it be held in the Quirinal Palace. The number of electors had been consistently over sixty since 1605, crowding the Vatican, and deaths of cardinals in recent conclaves emphasized the dangers of holding it there. The older cardinals refused to break with tradition, and the Vatican again housed the conclave. By the time it began on June 2, two cardinals had died, while four were absent. Ten of the thirty-four cardinals from Alexander's reign refused to follow Cardinal Chigi, whose lifestyle was disreputable. The sixteen from Urban VIII followed Antonio Barberini. The

Spanish party was only six strong, while the French faction had eight members. The eleven-man Flying Squadron's principal patron, Queen Christina of Sweden, was a most unusual person. She had abdicated her throne in 1654 and made her way to Rome the next year, where she made a highly publicized conversion to Catholicism. She had the unique honor for a woman of staying in the Vatican for a week when she first arrived in Rome. She became very close to Cardinal Decio Azzolino; it seems fair to say they were in love but were not lovers, despite the salacious rumors. Surely the best-educated woman of her time, Christina's views and quirky behavior had caused Alexander to become disillusioned with her and cut the large pension he was giving her. It was important to her to get a pope elected who both was sympathetic to her and met her standards of morality and virtue. The politically astute Azzolino served as her agent in the conclave and leader of the Flying Squadron.

The strongest *papabile* was Giulio Rospigliosi, who was acceptable to the Spanish, the French, and the Flying Squadron and Christina. French ambassador Chaulnes skillfully disguised his support for him by promoting Scipione d'Elci, allowing the Spanish to support Rospigliosi, well liked when nuncio in Madrid, although they would have preferred Francesco Barberini. Nonetheless, in the first two weeks Rospigliosi never received more than ten votes, and on the morning of June 20 his total was only five. Yet that evening he won the papacy with sixty-one votes. This dramatic turnabout shows how a talented diplomat could manipulate the process to the advantage of the king whom he served. The key in winning the tiara for the new Pope Clement IX was Chaulnes's promise of French revenues for Chigi, who brought his bloc over. The belief that both monarchs succeeded in getting a pope of their choice is made clear in post-conclave comments. Azzolino said to Christina, "This time the Holy Ghost wears a Spanish collar;" and Chaulnes wrote to his king, "Your Majesty makes the Pope in Rome as easily as the provost of merchants in Paris."[6]

The sixty-seven-year-old pope came from Tuscany. He was a theologian, but his reputation was built largely on being a fine musician and poet. He wrote operas that Alexander VII liked so well he gave the author a red hat in 1657. Clement's health was poor, which made observers wonder why the

French had not worked to get a healthier man elected. His two years as pope saw a reduction in the bitterness in the Bourbon-Habsburg rivalry. He provided modestly for his many relatives. One concern about him during the conclave was that he had many nephews; but only one, Jacopo, already an experienced diplomat, was made a cardinal. The Rospigliosi did rise in the ranks of Roman society through the marriage of a nephew into one of the great noble families.

Clement was generous to his friends in conferring red hats, promoting a Chigi, a Medici, and five others Tuscans out of twelve he named. He ignored the pressure from Catholic rulers to create crown cardinals and refused to name any non-Italians until the Turkish attack on Crete in 1669 forced him to give Spain and France a cardinal apiece to enlist their aid. Since the emperor was looking for help from him against the Turks in Hungary, Clement saw no need to name a German cardinal. He and his successors felt no obligation to name non-Italians who were not crown cardinals, but the rulers were equally opposed to appointments they did not control. Clement IX's health took a turn for the worse in October 1669, when news came of the fall of Crete to the Turks. He hung on until December 9. Fifty-four cardinals of the full complement of seventy were present when the conclave began on December 21, and twelve arrived late. The College was more divided than ever. Only eight of Clement IX's cardinals took part, giving Jacopo Rospigliosi less influence than Cardinal Chigi, who led Alexander VII's twenty-four cardinals. The French and Spanish parties were equal at eight, and the Flying Squadron had twelve, but it was not as cohesive as it had been in 1667. There was also a party of Romans, who wanted to elect a native of the Papal Sates. There was no clear favorite going into the conclave. Francesco Barberini, the dean, had worn the red hat so long that no cardinal had experienced his power as *cardinale-padrone,* but he still was known as a man with a fierce temper. Despite forty-six years as a cardinal, he participated in only five conclaves because of the long reigns during that time span. Pietro Vidoni, the choice of the Flying Squadron and Christina, was acceptable to Spain and France, but the delay in the arrival of the foreign cardinals required that his candidacy be keep under wraps until they came. Barberini opposed him because he had not been consulted about him.

France's clout in the conclave was made clear in the deference the cardinals showed to Louis XIV by marking time, despite bitter cold, until Ambassador Chaulnes and two French cardinals arrived on January 16. Even when Spain had been at its most powerful, the conclave had never been put on hold until its cardinals arrived. Spain had benefited in the elections a century earlier when France was wracked by religious wars; now it was at a disadvantage, because its ruler was eight-year-old Charles II. When Chaulnes arrived, he announced the exclusion of Elci, Chigi's candidate, with so much malice that Elci, deeply humiliated, fell ill and soon died. An effort to resurrect Vidoni's chances with strong French backing now met resistance from Chigi, angry at the French for excluding his choice. After the new chief minister took charge at Madrid, he sent a dispatch that arrived in mid-April declaring Spain did not exclude anyone, which freed the Spanish party to consider every candidate. An effort to elect Jacopo Rospigliosi reached thirty-three votes, at that point seven short of victory, as seven ill cardinals had left. It was the closest that a cardinal-nephew came to becoming pope after 1600. The next scrutiny saw voting return to the routine. This conclave saw an unusually large number of cardinals receiving votes; thirty-seven did on March 21, twenty-six with single ones.[7]

The factional lines slowly dissolved in late April, largely through the agency of Venice, and all the factions except the Flying Squadron agreed on one of Clement IX's creatures, Emilio Altieri. After the morning scrutiny on April 29 in which Altieri had three votes, a number he had received regularly from the start, they announced that he was their choice. In the afternoon scrutiny he received twenty-one votes and thirty-five by accession out of fifty-nine electors present. He was a month shy of eighty years, nearly deaf, and in poor health. He refused to leave his cell to accept the tiara, and two cardinals pulled him by his arms to the chapel to be acclaimed. Altieri's reluctance delayed an announcement to the crowd until midnight. Although she had opposed his election, Christina was the first through the reopened doors to do him honor.

The new pope became Clement X to honor the man who named him a cardinal only a month before the conclave opened. That remains the shortest time as a cardinal of any pope since Urban VI, the last pope not a cardinal. Clement was from a Roman family active in papal service; his deceased brother had been a cardinal. He had entered the church after a career as a lawyer. Having no nephews, he named as *cardinale-padrone* the uncle of his niece's husband, who was related to the Chigi. He was known as Cardinal Altieri. Since the pope was so old and feeble, Altieri had more power than any previous *cardinale-padrone*. He was anti-French, and his influence reduced the benefits that the French expected to gain from their role in Clement's election.

Clement X's reign saw more discord with Louis XIV, who asserted ever greater control over the French church. D'Estrées, the new French ambassador, put heavy pressure on the old man to name French cardinals, and in 1672 Clement promoted d'Estrées's brother and another Frenchman. He also added an Orsini and a Colonna, and those Roman families were again represented in the College. The French were dissatisfied with the promotions, and d'Estrées had an audience with the pope to complain that France had not been given due consideration. When Clement tired of his harangue, he began to rise out of his seat. D'Estrées pushed him back into it. The elderly pope shouted: "You are excommunicated!"[8] D'Estrées retreated in disgrace. Clement gained revenge by refusing to name any Frenchmen in his final promotion in May 1675, when the English exile Philip Howard was made the first English cardinal in eighty years.

Despite his age, Clement X kept an active schedule until 1676, when he began to fail. His horoscope had promised him nine years as pope, and his relatives were enraged when he died after six years. They blamed his physicians and assaulted them. The forty-four cardinals in Rome entered conclave on schedule on August 2, but d'Estrées demanded that they delay voting until the French cardinals arrived. The cardinals meekly acquiesced, and they marked time until the French reached Rome on August 29. With them the conclave included all sixty-seven cardinals. Only nine were not Italian, but both the French and the Spanish factions included enough Italians to total about twenty members each. The Flying Squadron numbered seven. The remaining cardinals were divided among the factions of the six cardinal-nephews present.

The presence of his brother in the conclave enhanced d'Estrées's control of the election. A wit among the cardinals joked, "The Holy Spirit used to be a dove; he has now become a cock!"[9] (*Gallus* is Latin for both cock and French.) Louis XIV made it clear that none of Clement X's creatures were acceptable because of Cardinal Altieri's anti-French attitude. Altieri wrote an abject letter begging forgiveness from Louis, who did not bother to reply. Altieri supported Benedetto Odescalchi, Innocent X's creation, who was the choice of the Flying Squadron. His strict neutrality allowed him to gain the votes of the Spanish party. In order to get French support for him, his supporters had to resort to a subterfuge. Rospigliosi and Chigi wrote to Louis claiming that Altieri's support for Odescalchi was a ploy designed to get the Flying Squadron's votes for his and Christina's true favorite, Cardinal Conti. Louis fell for the stratagem and wrote d'Estrées that Odescalchi was satisfactory. It took a week after his letter reached Rome before Odescalchi was elected on September 21, 1676 with sixty-one votes. Two cardinals had died by then, and three had left the conclave in ill health.

Odescalchi refused to accept the papacy until the cardinals agreed to swear to an election capitulation that he had drawn up at the previous conclave. What made this capitulation unique was that the new pope required the cardinals to swear to observe it. Its major points included reform of the clergy, independence of the Church from the control of secular rulers, and a restoration of the cardinals' traditional role as advisers to the pope. Once the cardinals agreed to it, he announced that he would be Innocent XI, in honor of the pope who had made him a cardinal in 1645. The thirty-one years he had been a cardinal contrasted sharply with Clement X's one month. Born at Como in 1611, Innocent was trained in law. He had risen rapidly in the hierarchy, but after becoming archbishop of Novara in 1650 he resigned because of ill health. It had been a factor in keeping him from the tiara in the past two conclaves, but his health had much improved by 1676, and he reigned for thirteen years. He governed the Church without a *cardinale-padrone,* for he was fiercely opposed to nepotism. Innocent's nephew Livio received no major office, nor were other relatives enriched. "May you fare like Livio Odescalchi!" became a way for the Romans to wish others bad luck.

The ever more bitter struggle between pope and French monarch was the major concern of Innocent XI's reign. It centered on *régale,* the right of the French king to take the revenues from the dioceses of dead bishops until their successors were installed. The popes had been seeking to abolish it since Trent. Innocent, who had an exalted sense of papal authority, made a concerted effort to eliminate it, while Louis XIV was not one to take kindly to any reduction in his authority. Innocent's bull condemning the practice led to a national synod of French clergy in 1682. It produced the "Four Gallican Articles," a statement on the rights of the French church vis-à-vis the papacy, including the right of *régale*. The pope responded by refusing to confirm new French bishops, and by 1688 there were thirty-five vacant French dioceses. Louis in turn occupied the papal territory of Avignon. Even Louis's revocation in 1685 of the Edict of Nantes, ending toleration for the Huguenots, failed to smooth over relations with Rome.

Of course, such matters affected the pattern of appointing cardinals. Innocent XI, who canonized few saints, because few souls met his criteria for sainthood, felt the same way about cardinals. He was so slow about making his first promotions that many believed he intended to reduce the College to fifty. His first appointments came in 1681, when he gave out sixteen red hats. All the new cardinals were Italians, which angered the Catholic rulers, since most non-Italians in the College had died. Innocent's second creation of cardinals occurred five years later, when he gave out red hats to fifteen Italians and twelve non-Italians, including only one Frenchman. When Innocent XI died on August 12, 1689, he left eight vacancies in the College. Steps were soon taken to canonize him, but the French objected, regarding him as too hostile. In 1956 Pius XII beatified him, but he has not yet been canonized.

Nine of the sixty-two cardinals did not reach the conclave. Seven participants were non-Italian, while seventeen came from the Papal States. Innocent had promoted twenty-seven; the rest were about equally divided among the four previous popes. Innocent's refusal to name a cardinal-nephew meant that his cardinals lacked cohesion and did not take advantage of their numerical strength. Innocent had largely succeeded in appointing politically neutral cardinals, as the French party consisted of only five members, and the Habsburg, just seven. More influential was a group of nine cardinals calling themselves the *zelanti,* whose goal was to elect the best pope regardless of political ties. The Flying Squadron, which by then was regarded as simply another political faction, had largely dissipated upon Azzolino's and

Christina's deaths earlier in 1689. The cardinals entered conclave on August 23, but again they were forced to delay any meaningful balloting until the French cardinals arrived on September 23. It was clear from the first that Pietro Ottoboni was a strong candidate, partly because he was seventy-nine years old. After his conclavist promised that he would accept the unconfirmed French bishops, the only obstacle to his election was removed. On October 6, 1689, the forty-nine cardinals still in conclave unanimously elected Ottoboni pope. He chose the name Alexander VIII to honor the pope who started his rise in the hierarchy, although it was Innocent X who had given him a red hat in 1652. He came from a patrician family of Venice and had studied law. He was known to enjoy spending his revenues, in contrast to his frugal predecessor.

Alexander VIII also differed from Innocent XI in reviving nepotism. He named his twenty-year-old grand-nephew Pietro *cardinale-padrone* and filled papal offices with relatives. He allegedly said that he needed to make haste since he was already in the twenty-third hour. Alexander began to implement the promises made for him to Louis XIV during the conclave, but when it became clear that Louis was not reciprocating, he stalled on fulfilling the rest. With the War of the League of Augsburg (1688–1713) raging, in which France stood against most of Europe, Louis dared not alienate the pope too badly. The issues that had so roiled the previous reign were left on the back burner. As a Venetian, Alexander was less generous to Austria, because Venice regarded Austrian successes against the Turks as threatening. The political change was reflected in the red hats Alexander distributed in February 1690. The only non-Italian among the ten new cardinals was a Frenchman for whom Louis had petitioned the late pope. The Habsburg cardinals refused to attend the consistory in which his red hat was granted. Habsburg anger increased when two friends of Alexander's nephew were promoted late in 1690. The Austrian ambassador was recalled from Rome, and the Imperial cardinals were told not to participate in curia business, but to stay in Rome for the next conclave.

Yet Alexander's last act before he died on February 1, 1691, condemned the Gallican Articles of the 1682 French synod for infringing on papal rights. That raised the stakes when the conclave began on February 12. The

College was full, but nine cardinals did not attend. The French reached Rome on March 27, the longest delay yet. It was clear early on that Gregorio Barbarigo, a *zelante,* was a strong candidate. France supported him, since he was a Venetian, but Leopold I of Austria ordered his cardinals to prevent his election for the same reason. Leopold preferred not to use a formal exclusion if possible, but word spread that Barbarigo had been excluded. The *zelanti* demanded that his election go forward, but enough cardinals accepted the right of exclusion to prevent it. The fact that the worldly cardinals thought he would be another Innocent XI was also a factor in keeping him from the tiara.

By late April it was obvious that Barbarigo would not be pope, and the conclave became directionless. Scrutiny after scrutiny failed to provide any hint of who might win. Votes went to noncardinals, which had not happened since 1503. The tedium of such long conclaves must have been severe, as ballot after ballot failed to show any change in the pattern of voting. Often the morning and afternoon scrutinies would be identical. Overnight discussion would result in a few changes the next day, but for weeks on end there was no hint of how and when the conclave would end. The most excitement came from the fire that broke out in the Vatican late one night in June, after a lamp was knocked over by a cardinal who was playing cards. It ruined several cells, but the deaths of three cardinals previously had opened enough space for those who were burned out.

Cardinal Altieri, now an elderly statesman, began an effort to get himself elected. He had worked hard to ingratiate himself with Louis XIV but had maintained a public image of being pro-Habsburg. The *zelanti* and Chigi opposed him, and they carried weight with enough other cardinals to keep Altieri from victory. His strong showing, however, gave him credibility as a faction leader he had previously lacked, and once it became clear that he would not be pope, he set to work to get his friend Antonio Pignatelli elected. A boomlet in his favor in March had reached a majority before falling back. In late June, as the heat mounted and more cardinals took ill, his candidacy was resurrected. Still it was July 12, 1691, before Pignatelli reached the required two-thirds majority, largely because Altieri had to work hard to convince the French faction that despite being from Naples, he would not be a tool of the Spanish. Most *zelanti* stubbornly refused to vote for him, although he was Innocent XI's creation, as they mostly were. Out of gratitude to that pope, he became Innocent XII.

Innocent was seventy-six years old and trained in law. Soon after his election he took a bad fall that left him unable to walk for a year, but he was in generally good health during his nine-year reign. He is most noteworthy for issuing a bull in 1692 banning papal nepotism. To be sure, it was easy for him to do, because he had no nephews. Leading up to it was a study of the sums and offices given to papal nephews since Urban VIII, who was shown to be the worst offender. Urban's nephews had received 1,700,000 *scudi*, while Alexander VII's were not far behind at 1,400,000. The bull barred popes from enriching their relatives in any way from Church property. Only one papal relative could be given a red hat, should there be one worthy of it, and his revenues were limited to 12,000 *scudi*. It eliminated several offices in the Papal States that often went to the popes' lay relatives. Innocent required that all cardinals at future conclaves and every new pope swear to obey the bull. Although there were several breaches in the future, the bull eliminated papal nepotism. It marked the rise of a centralized administration with professional bureaucrats with careers in papal service, which served far more effectively in seeing that the pope's decisions were carried out and greatly reduced the need for papal nephews. The most obvious element of the new approach was filling the office of secretary of state with a professional diplomat instead of the cardinal-nephew.

Innocent waited until 1695 to create his first cardinals, with a Pole as the sole non-Italian among the ten promoted. In later consistories, he was more generous to the rulers, promoting a prelate from Portugal and two from both France and Spain. Emperor Leopold I had to be satisfied with red hats for two Italian nuncios at his court, and a nuncio to Switzerland received one, the first time it had figured in the College in any fashion since 1510. Innocent proposed building a new hall for housing the conclave, but nothing came of the idea. He resolved the long-standing dispute with Louis XIV over *régale* by confirming the bishops whom the king had appointed since 1682, while Louis agreed not to require his new bishops to swear to uphold the Gallican Articles of 1682. The pope gave Louis a far greater prize in 1700 when he recommended to childless Charles II of Spain that he recognize

Louis's second grandson Philip as heir to his throne over Leopold's younger son. Both princes had a claim to the Spanish succession as offspring of Charles's sisters. Charles's will granted the Spanish throne to Philip on the condition that he renounced any claim to the French crown. Charles died six weeks after Innocent did on September 27, 1700. Their last major decision led to the War of the Spanish Succession (1702–13).

The conclave that opened on October 9 with fifty-eight of sixty-six cardinals present was fraught with tension over the impending war. The Spanish cardinals were left directionless because of the deathwatch on Charles II, and they no longer worked with the Imperial party. The *zelanti* were far more numerous than in 1691, totaling thirty-one, including eighteen from Innocent XII's reign. The remaining sixteen were open to persuasion by any party. Since six of the seven French cardinals were in Rome for the jubilee of 1700, the long delay before serious balloting began of the past conclaves was not repeated. Since no one received even half of the votes, however, the conclave dragged on into November.

The arrival of the last French cardinal on November 14 with Louis's instructions had little impact, but three days later the news of Charles II's death transformed the conclave. The obvious need for a nonpolitical pope meant that a *zelante* would be elected. Giovanni Albani, author of Innocent's bull on nepotism, quickly emerged in front. At first the French faction resisted but soon fell in line, and on November 23, Albani was elected unanimously. He was reluctant to accept the office, it was said, because he had ambitious nephews who would be angry if he enforced the bull against nepotism. He agreed only after consulting theologians, who argued that refusing a unanimous election was resisting the Holy Spirit. He chose the name Clement XI, because he was elected on the feast of St. Clement. Born at Urbino into a noble family, the new pope was fifty-one years old, a youth compared to the prior four popes. Alexander VIII had named him a cardinal-deacon in 1690, but he was ordained a priest only weeks before the conclave began, presumably to make him a more appealing candidate. His coronation had to wait for his episcopal ordination.

The major issue Clement faced was the Spanish succession. Louis XIV had accepted Charles II's will naming Philip of Bourbon as ruler of the undivided

Spanish empire, which gave him control of Naples and Milan, and the pope quickly acknowledged him as Philip V. But Louis overplayed his hand by renouncing the clause that prohibited the union of the French and Spanish crowns, and the other major states of Europe joined with Emperor Leopold in demanding that his son Charles be given the Spanish crown. When both sides rejected Clement's offer of mediation, war broke out in 1702. The pope's position swung rapidly from neutral to pro-French. Austria now saw the papacy as an enemy and invaded the Papal States in 1706, as its army marched toward Naples. Under duress Clement recognized Austrian rule there only to have Philip break off relations with Rome in protest. For twelve years Spain was in schism. But when the Austrian choice to become Spanish king became Emperor Charles VI in 1711, his allies would no longer support his claim to Spain. The Peace of Utrecht of 1713 left Philip V on the Spanish throne, while Naples and Milan passed to Austria, allowing the Austrians to dominate Italy.

In 1715 Louis XIV died, leaving a great-grandson as Louis XV. The dead king's younger brother, who was less rigid on royal rights over the French church, became regent. Papal relations with France improved, which was reflected in the red hats given out late in Clement's reign. Among his first twenty cardinals, eighteen were Italian. In his later years he added fifty men to the College, and the non-Italians numbered fifteen. France gained six, while Hungary gained its first cardinal in a century, and Bohemia, its first since 1400. Nuncios dominated the Italian contingent, making up nearly half. Clement promoted his nephew Annibale but gave him little power.

Clement had fallen ill in 1710 and never again was in good health. He spent his time at Castel Gandolfo, the papal summer residence, where he spent lavishly for buildings and gardens. He died on March 19, 1721, after a reign of twenty years, the twelfth longest among popes. Before going into conclave the cardinals had to decide whether to admit two cardinals whom Clement had barred. Giulio Alberoni had negotiated a marriage between Philip V and Elizabeth Farnese of Parma in 1711 against Clement's orders, and the royal couple had browbeaten him into giving Alberoni a red hat. When he lost influence at Madrid in 1720, the pope gained revenge by stripping him of his privileges, including the right to vote for pope, while Louis de Noailles's defense of the Gallican Articles had so exasperated Clement that he disenfranchised him. The College reaffirmed that a duly constituted cardinal did not lose his vote except for apostasy, and invited both to the conclave, but Noailles did not attend.

There were two vacancies in the College, and twelve cardinals were not at the conclave that began on March 31 with only twenty-seven present. By the end the number was fifty-five, one cardinal dying during it. Ten absentees were Clement's creatures, further reducing the clout of his nephew, who was not capable of managing a conclave party. The French monarchy sent gold to Cardinal Albani, demonstrating that bribery had not disappeared from conclaves. This point was further proven by the emperor's gift of a priceless diamond to Albani. A delay in its reaching him allegedly explained why he did not support a pro-Austrian candidate early in the conclave. The College was divided into four factions, whose blocs were more fluid than in recent elections. Albani could count on about sixteen of his uncle's men; the *zelanti* numbered eight; the Franco-Hispanic faction had fifteen by the end; and the Habsburg party was about the same size. The last had an advantage early on, because most of its members were already in Rome, while the Franco-Hispanic cardinals arrived well after the first ballots. It was clear that age would be a major consideration; the cardinals did not want another twenty-year reign. A respected cardinal was deemed too young at sixty-one.

At the first scrutiny on April 1, 1721, Albani asked his uncle's cardinals to show their gratitude by voting for Fabrizio Paolucci, Clement's secretary of state. He came within three votes of being elected. Surprised by the vote, Michael von Althan, the Habsburg leader, had to announce that the emperor had excluded him as anti-Austrian. Althan spoke directly to each cardinal to repeat the veto. This secular interference deeply offended the *zelanti,* who continued to vote for Paolucci. Armand de Rohan, the Franco-Hispanic leader, insisted on the right of a ruler to exclude anyone, and this led to a rapprochement between the two political factions. They lacked enough votes to get any of their choices elected until more cardinals arrived, and the conclave fell into marking time. As Albani and the *zelanti* promoted various *papabili,* the political leaders revealed further exclusions from their governments. By April 25, after all the participating cardinals had arrived, the possibilities were so far reduced that aged Cardinal Pignatelli, who had remained away because of ill health, was persuaded to hobble into the Vatican for consideration. His sacrifice failed to pay off, as Spain excluded him.

The political leaders turned to Michelangelo dei Conti, who made extensive promises to both political factions. He was one of Clement's creatures, but Albani was not well disposed to him, nor were the *zelanti*. Long negotiation overcame their objections, and on May 10 he became Innocent XIII by unanimous vote. The new pope came from the same Roman family as Innocent III, whose name he took. He was born in 1655 and had a career in papal diplomatic service. Named a cardinal in 1706, he had long suffered from gallstones. He named only three cardinals—his brother Bernardo, Albani's nephew Alessandro, and a Venetian. Alessandro Albani was a bon vivant about whom it was joked that he must have been born on a Sunday, because he never did a day's work.

Innocent was seriously ill for a year before he died on March 7, 1724. Preparations for the next conclave began long before then, yet twelve cardinals did not attend. Neither the factions nor the *papabili* had changed much since 1721. The conclave began on March 20 with thirty-three cardinals present; their number increased to fifty-three by the end. Annibale Albani again supported Paolucci, and much the same intrigue as in 1720 was played out, with Albani trying to secure his election before a definitive exclusion arrived from Vienna. Again some voters insisted on casting ballots for him in protest against such exclusions, and the conclave dithered into May as cardinals arrived one or two at a time. For the first time in two centuries, English influence played a role, albeit minor, in a conclave. The English agent in Rome, offended by the honors accorded to the exiled Stuarts at the papal court, invested money in electing a pope who would oust them from Rome or at least reduce their honors. He found an accomplice in Alberoni, but his influence was too slight to achieve his goal.

The Imperial party promoted Giulio Piazza, who came within four votes of election on May 13, and with more members arriving soon, it had good reason to expect victory. The word about Rome was that Piazza would be the new pope. Albani, greatly annoyed at being left out of the negotiations over him, derailed his cause. He proposed Pietro Orsini as an alternative. Orsini's lack of political experience had deterred support for him, but now his band-wagon was rolling, and all the cardinals eventually declared for him. The ar-gument that the Holy Spirit had spoken overcame his reluctance to accept the office. Persuading him took two days, and it was only on May 28 that he con-sented. He wanted to call himself Benedict XIV, thereby recognizing the last Avignon pope, but curia lawyers convinced him to become Benedict XIII.

The new pope, age seventy-five, was a member of the powerful Roman family that had provided two popes in the Middle Ages. After joining the Dominicans, his family's influence gained him a red hat at twenty-three, but there had been little objection to his youth, because he already had a repu-tation for piety. His otherworldliness had kept him from being a serious can-didate for the papacy in past conclaves, and he changed little on becoming pope. He left church administration to Niccolo Coscia, a commoner who had befriended him when he was a young priest. It was Coscia serving as his conclavist who engaged in the intrigue needed to get him elected. The pope made him a cardinal and handed over power to him. Benedict loved the city of Benevento, where he had been archbishop, and he kept the see as pope, being the last to have two dioceses. The major issues he faced involved church affairs. As a Dominican he distrusted the Jesuits and began to wean them from the influence they had enjoyed at the papal court since 1560. He did little for his family, but Coscia filled papal offices with his own relatives and cronies from Benevento, where he had served as chancellor. Coscia was hated because of his fiscal policies. When Benedict died in February 1730 from pneumonia caught at a cardinal's funeral, Coscia fled Rome in disguise while the people sacked his house.

Benedict XIII had created twenty-nine cardinals in twelve consistories, naming two or three at a time instead of many at once. Each of the five major Catholic states had a crown cardinal appointed. Portugal demanded that its king's son receive one, but the pope refused unless he became a priest, which he never did. Benedict's stand was a key step in freeing the Col-lege from such royal-blooded cardinals. He also refused to name his nephew, although the cardinals asked him to do it.

When the cardinals met in congregation after Benedict's death, their first task was clearing out Coscia's cronies and relatives from papal offices. They begrudgingly allowed him into the conclave. Sneaking back into Rome, he entered the Vatican soon after the conclave began on March 3, but he remained in his cell. Thus, Benedict's twenty-two cardinals present lacked a clear leader, while Annibale Albani led Clement XI's twenty-five creations. Eight cardinals from three other popes filled out the fifty-five participants. Soon after the conclave began with twelve absentees, elderly Bernardo Conti was a victim of a practical joke. Albani told him that his election was all set, and he became so excited that he collapsed and died, presumably from a heart attack.

In addition to factionalism determined by papal patronage, there were again the *zelanti,* who cut across those lines. The major issue was the fate of Tuscany, where childless Duke Giovanni was near death. His brother Francesco, last of ten Medici cardinals, had given up his red hat and married in hope of keeping the dynasty going, but he too was childless. The next pope would influence the choice of the new duke, which made this election a serious concern for all parties. With many non-Italians still on the road, the *zelanti* promoted eighty-year-old Lorenzo Imperiali, who had been strong in four previous elections. The first scrutiny on March 6 gave him eighteen votes, six short of victory. Several among those who entered conclave in the next days supported him, and he reached twenty-two votes, now five short of victory. Then the Austrians presented a veto of Imperiali and ended his candidacy.

As had happened in the previous elections, the conclave now marked time until all those on the road arrived, which only happened on April 17. That was also the day on which someone voted for Coscia, which created tumult among the cardinals. Those who argued that any duly constituted cardinal had a right not only to vote but also to receive votes prevailed over those who wanted to declare the ballot invalid. The matter of the Tuscan succession now intruded into the conclave. The Spanish crown announced that it had determined that a Spanish prince was the duchy's rightful heir. The uproar in Rome was enormous, and the conclave swung toward finding a pope who would prevent such aggrandizement of Spanish power. Soon attention became focused on the *zelante* Lorenzo Corsini, a Florentine who was regarded as neutral in his politics and the right man to do what was best in respect to Tuscany. His nephew Neri, a fine diplomat, aided him. Probably more to his advantage, Corsini was seventy-eight years old. It took two more months to elect him, as the leaders of the political factions had to advise their courts and wait for word on him. On July 12, after being in conclave for 131 days, the dean announced to the weary Romans the unanimous election of Clement XII, a name Corsini chose to honor the pope who had given him his red hat in 1706.

From a wealthy merchant family with several cardinals in its past, the new pope had entered the clergy after a career in diplomacy. Despite his age and severe gout, Clement began his reign with several initiatives. He convicted Coscia of malfeasance and imprisoned him, issued paper money for the Papal States, and condemned freemasonry. Within a month of his election he named his nephew Neri a cardinal. While Clement's family was so wealthy that Neri did not need lucrative offices, the pope's poor health handed him immense power, and he probably had more authority than any other *cardinale-padrone*. In a year the pope went blind and was bedridden for the rest of his ten-year reign. His mind remained sharp, and he had the final say in major decisions.

Clement's reign saw the Spanish regain the kingdom of Naples in 1735, and their threat to Rome forced him to accept Carlos of Bourbon as its king. When the duke of Tuscany finally died in 1737, Clement's anger at the Spanish motivated him to broker a deal by which Duke Francis of Lorraine, husband of the future Austrian empress Maria Theresa, became the new duke. Difficulties with the other Catholic rulers led the pope to refuse to name any non-Italian cardinals during his first seven years. Twenty Italians were appointed before he named a crown cardinal for each of the five major Catholic realms. Most notoriously he gave a red hat to a nine-year-old son of Charles III of Spain, the youngest cardinal ever, who later gave it up to marry. Clement appointed a total of seven non-Italians out of thirty-three creations. Despite ill health he hung on until February 6, 1740, dying days short of turning eighty-eight.

Clement XII's long-time-a-dying meant that the *prattiche* had been underway long before the conclave began on February 19 with thirty-two of sixty-eight cardinals present; the participants reached fifty-six by its end. Ten absentees were non-Italians, while two Italians were too ill to attend. Imprisoned Cardinal

Coscia successfully petitioned the College for entry. The conclave's politics were unusually complicated. Each of the five major Catholic powers had something of a party; the *zelanti* numbered about ten; and the creations of Popes Clement XI and Clement XII counted sixteen and thirty respectively, although not all adhered to their patron's faction. Alexander VIII's nineteen cardinals did not form a coherent faction. Albani and Corsini, leaders of the main papal factions, were already hostile to each other, and it grew worse as the conclave dragged on. The struggle between them to get their choice elected dominated the election, as the royal parties lined up with one or the other. This conclave revealed a striking case of the naïveté often displayed by non-Italian cardinals. French Cardinal Rohan arrived in Rome proclaiming that he had come to make Cardinal Olivieri pope. Not only did that ruin Olivieri's chances of being elected, but it also convinced him that Rohan for some unknown reason had acted out of malice toward him.

By 1740 the Catholic rulers understood the diplomatic cost of an open veto and avoided using it in this conclave. Emperor Charles VI, however, was determined to prevent the election of a cardinal from Naples, and his faction leader waged a quiet but effective campaign against the three Neapolitans, including Tommaso Ruffo, regarded as the strongest candidate going into the conclave. When he repeatedly failed to muster two final votes to win, the factions moved on to other *papabili*. Corsini then promoted Leander Porzia, one of Benedict XIII's creatures, who had passionately addressed the cardinals on the need to conclude their business. He impressed enough electors to come within one vote of election on April 9. Holy Week halted balloting, although in some conclaves voting was done even on Easter. Before it resumed, a broadside lampooning Porzia was found in the Vatican. He erupted in anger so fierce that he suffered a stroke and died, it was said, of "papal rabies." Who brought the broadside in was never discovered. Porzia was one of four cardinals to die during the conclave and a fifth withdrew for ill health.

The skilled diplomat Corsini soon had another man on the verge of election. Pompeo Aldrovrandi was only one vote short of victory on July 9, but Albani was opposed to him and was able to prevent any of his faction from defecting to him. For a month Corsini continued to push Aldrovrandi, who constantly was one or two votes shy of election. Finally he asked his voters to support others. The conclave fell back into confusion, but now the *zelanti* asserted themselves after playing only a negative role. They began to canvass for Prospero Lambertini, kin to Aldrovrandi. As Benedict XIII's creature, he had fallen outside of the purview of the factions despite being well respected. He had been receiving a few scattered votes, and insisted later that he made no effort to win votes. Suddenly all fell into place for him. Following a ballot where he had received nary a vote, he was elected unanimously in the 255th scrutiny, on August 17, 1740.[10] It was the most dramatic turnabout in history, while the six-month duration of this conclave remains the longest since the 1316 conclave of twenty-eight months.

Benedict XIV came from a minor noble family of Bologna. Born in 1675 and educated in law and theology, he received his red hat in 1728. He was already known for his learning, spirituality, sense of humor, and ability as an administrator. That it took six months to elect a man who would prove to be a great pope says much about the nature of papal elections in that era. Benedict set up concordats with five realms, which lessened papal influence over the clergy into those lands but reduced major sources of controversy.

In church affairs, Benedict is recognized for writing the first papal encyclical, which dealt with the duties of bishops. He reduced the severity of the Index of Forbidden Books and the barriers for a Catholic marrying a non-Catholic, but he also issued the definitive ban on the Chinese rites, in which the Jesuits had incorporated elements of Chinese culture into Catholic services. This ruined a successful missionary enterprise in China. It reflected a growing suspicion of the Jesuits within and without the Church that erupted into a major controversy in the next reign.

Unimpressed with the possible choices for red hats, Benedict delayed giving any until late 1743, when he appointed six nuncios and nineteen other Italians along with a Bavarian and a Knight of Malta. The absence of crown cardinals enraged the Catholic powers, and the pope had to tolerate their protests for three years until enough vacancies opened up to accommodate all five of them. In 1747 Benedict promoted Duke Henry of York, the last of the Stuarts, whose sixty years as a cardinal are the most ever.

Benedict XIV had an amazing reputation among Enlightenment philosophers for his learning. When he died on May 3, 1758, Voltaire was among those who expressed regret. As the conclave began on May 15, it was clear that the future of the Jesuits was the major issue, since Benedict had given authority to investigate the order to Portuguese minister Pombel, whose goal was its destruction. The forty-five cardinals who attended were divided between Jesuit supporters and opponents. The latter, mostly Benedict's thirty-four creations, lacked an effective leader, while Neri Corsini led the eleven from earlier popes, giving them more clout than their number warranted. France and Portugal opposed anyone friendly to the Jesuits.

Again the delay in the arrival of the non-Italians bringing instructions from their courts meant that little was done until mid-June. Carlo Cavalchini, Corsini's candidate, reached twenty-eight votes on June 22, two short of victory, and he was expected to pick up five more at the next scrutiny. The leader of the French party was forced to produce his king's veto of Cavalchini. Corsini protested: "You French must always contradict the Holy Spirit!"[11] Cavalchini for his part fell on his knees and thanked God for sparing him the burden of papal office. Other candidates were then promoted and failed, leaving Carlo Rezzonico as the only *papabile* still untested. As a Venetian he had opponents, but when a dying cardinal was removed during a heat wave, the cardinals decided that they needed to finish their business. On July 6, Rezzonico was elected with thirty-one votes, the lowest percentage in over two centuries.

The new Clement XIII, honoring the pope who named him a cardinal in 1737, was sixty-four years old. He soon may have wished that he had been more steadfast in his initial refusal to accept the tiara, for within days of his election the matter of the Society of Jesus came to a head in Portugal, where Pombal dissolved the order and imprisoned or expelled its members. Over the next ten years the Jesuits were ousted from France, Spain, Naples, and Milan. The pope objected strenuously, but none of the kings backed down. In 1768 three rulers demanded the order's dissolution. Clement's anguish over the matter ruined his health, and he died on February 2, 1769, a day before he was to meet with the cardinals to formulate a reply.

Clement's sixty-two appointments to the College were mostly Italian, including his nephew, who, all agreed, deserved a red hat. The five Catholic powers had received one crown cardinal each. Many of Clement's creations died before he did, so there were fourteen vacancies when the conclave began on February 15. Only twenty-eight were locked into the Vatican Palace that day, but forty-six eventually participated. The issue of the Jesuits's suppression dominated the election. The anti-Jesuit powers were adamant about excluding anyone they regarded as lax on the Jesuits. The Spanish ambassador said that a general war in Europe would be less harmful than a pope who favored the Society. The cardinals favoring the Society, mostly the *zelanti*, were eager to get a pope elected before the French and Iberian cardinals arrived, but the anti-Jesuit faction announced that it would exclude any candidate they proposed. The coming of the non-Italians took longer than ever; the Spanish entered on April 30. They had set out by sea, but storms forced them back, so they came by land, which took three weeks longer.

Besides seeking ways to entertain themselves, the cardinals in conclave meanwhile debated the merits of the *papabili*, largely determining how they stood on the Jesuits. The ambassadors provided lists of who was acceptable and who was not, but the lists contradicted each other and together left only about ten cardinals. A Spanish proposal that the cardinals be required to sign a pledge to suppress the Society in exchange for support from the anti-Jesuit faction was denounced as simony. In March the visit of Joseph II, coemperor of Austria with his mother Maria Theresa, broke the routine for the cardinals. He entered the Vatican and dined with them, the only outsider ever thus honored; they even permitted him to keep his sword on him. He provided little direction on the election, however, being indifferent on the Jesuits.

When the final arrivals appeared, each non-excluded candidate was promoted in turn, but the twelve *zelanti* were adamant against interference by the secular powers, and they could count on additional votes against each candidate on other grounds. After going through the list, the cardinals returned to Lorenzo Ganganelli, who had reached twenty votes early on. Many thought him unreliable, but rejecting a Spanish demand that he vow to suppress the Jesuits rebounded in his favor. The crown cardinals had to respect papal freedom of action, even as they sounded out Ganganelli for his views on the Society. He said that he knew of no canonical objections to dissolving it. A night meeting on May 18 between the crown cardinals and the *zelanti* led to an agreement on Ganganelli. The next morning he was elected unanimously. The rulers amply rewarded the crown cardinals for securing his election, but as long as nothing had been promised beforehand, such rewards were not deemed simony. An agent for the city of Lucca produced an account of the conclave's cost. He estimated that Clement XIII's funeral and building the conclave cells cost 70,000 *scudi,* while the conclave's daily expenses were over 20,000, for a total of above 1,930,000 *scudi.* Ten years earlier, total annual revenues from the Papal States were estimated at between two and three million *scudi.*[12]

Ganganelli became Clement XIV to honor the pope who named him cardinal in 1759. He was born near Rimini in 1705 and became a Franciscan. His red hat was largely a result of his closeness to the Jesuits, as Clement XIII wished to balance the anti-Jesuit crown cardinals. As a cardinal, however, Ganganelli bent to the prevailing wind and angered his patron with his new attitude. As a result he gained little experience in the curia. Instead of becoming dependent on a nephew or some other confidant, the new pope became secretive and conducted papal business with little advice from the cardinals. He was equally secretive in giving out red hats. He made eleven promotions *in petto* and died without revealing them. Most of his seventeen published cardinals were Italians whom Clement knew from the congregation he headed as a cardinal. He left fifteen vacancies in the College at his death in September 1774.

On the crucial issue of his reign, the suppression of the Jesuits, Clement vacillated at first, but finally in 1773 he issued a bull dissolving the order for disturbing the peace of the Church. In turn the king of Naples restored two duchies to the papacy. That the return of land to the Papal States should be the reward for suppressing an order so valuable in enhancing papal authority, demonstrated yet again the enormous liability that sovereignty over them imposed on the papacy. The suppression was also a result of a papal electoral system that gave veto power to the major Catholic powers. The situation had become worse when the Bourbons gained Naples in 1735. In short order the Neapolitan cardinals had become Bourbon crown cardinals and added substantially to Bourbon clout, which allowed the Bourbons to dominate the conclaves until 1789.

CHAPTER IX

Conclaves in the Nineteenth Century

In the seventeenth and eighteenth centuries, the conclave was the most uniform in format, procedure, and results that it has been throughout its history. Every conclave took place in the Vatican palace with about sixty cardinals attending, of whom 90 percent were Italian. They each lasted about two months, and the elections were controlled by the dominant political power of the era, usually France. The popes chosen in these conclaves conformed closely in social status, careers, and age. Just as the social and political structures of ancien régime Europe were deemed permanent and ordained by God, so too the method of choosing the pope and the type of man chosen were seen as having reached their final divine perfection. The coming of the French Revolution would shatter all of those illusions.

Appropriately, the conclave that elected the pope who would experience the terror of the French Revolution was the most routine of the eighteenth century. The conclave began on time after Clement XIV's death on September 22, 1774. Clement left fifteen vacancies in the College, and there were eleven cardinals absent. Cardinal de Bernis, the French leader, was the sole non-Italian among twenty-eight cardinals present at the start. He persuaded the College to wait for the other crown cardinals, although only three made the trek to Rome. Aware that the election would not happen without them, they dithered on the way, and it took seventy days for the last to arrive, so serious balloting only began on December 10. Unlike many conclaves where nothing interrupted the process, no balloting took place on

Christmas. A satire on the cardinals' foibles penned by a priest of Florence circulated in Rome and reached the Vatican. It outraged the cardinals, who ordered the Inquisition to destroy every copy and banish the author.

The principal issue was still the Society of Jesus. About half of the cardinals wanted to restore it, while the crown cardinals and some Italians wanted a pope who would enforce its suppression. Gianangelo Braschi was the strongest *papabile* going into the election, but the Austrians declared him unacceptable. Braschi was four votes short of victory before Christmas, and no one else came close. For two months others were unsuccessfully promoted. Charming and affable, Braschi's appeal lay in that both the pro- and anti-Jesuit factions had reason to believe that he was sympathetic to their sides. He was prevailed upon to say that he would never restore the Society, but its supporters remained convinced that he would do otherwise. Both sides worked to persuade the Austrian leader to say that he had not vetoed Braschi but only recommended against him. When it was announced that Austria had not formally excluded him, he was elected unanimously in the 265th scrutiny on February 15, 1775. He became Pius VI to honor Pius V.[1]

Pius was fifty-seven years old and vigorous. The Braschi probably were Swedish originally, but they had been living in the Romagna for over two centuries. Pius was educated in law. When his fiancée jilted him, he joined the clergy. He was ordained a priest only at age forty-one and was not yet a bishop at his election. He had served as a conclavist in the election of 1740 and became a cardinal a year before his election. Were it not for Pius's fate of being pope during the French Revolution, he might today be known for resurrecting the old evil of papal nepotism. His nephews Romoaldo, named a cardinal, and Luigi received papal offices and vast incomes. The term *cardinale-padrone* was again used for Romoaldo, but Luigi offended more with his avarice and arrogance. Nepotism motivated one of Pius's more worthy projects, draining the Pontine marshes south of Rome, as Luigi received rent from most of the land cleared.

Pius was the first pope to leave Italy since Clement VII went to Marseille in 1533. In 1782 he traveled to Vienna to confront Emperor Joseph II over his policy of tolerating Protestants and Jews, closing monasteries, and restricting papal control over the Austrian church. Although the trip was a triumphal progress, Pius failed to convince Joseph to change his policy. The pope also failed to convince the Prussian and Russian rulers to stop protecting Jesuits in their realms. His policy toward the suppressed order was ambiguous, doing nothing to reverse the bull of suppression but not bothering those Jesuits who continued to live a common life.

At Pius's election the College was short eighteen members. Between 1775 and 1789 he created fifty-four cardinals; among the twelve non-Italians was the last cardinal to serve in the French government. On the eve of the Revolution, Pius made his last major set of promotions, adding four crown cardinals and five Italians to the College. In the last ten years of his reign he named eleven cardinals, but two were *in petto* and never revealed. In all Pius VI created seventy-three cardinals; 78 percent were Italians, nearly the norm for the previous two centuries.

The first fifteen years of Pius's reign were largely peaceful in Europe. The American War of Independence eventually involved nearly all the major powers, but it had little direct impact in Europe and none on the Papal States. Pius had little understanding of the issues involved in the American Revolution, but he took the opportunity offered by American independence to create the first diocese at Baltimore, with John Carroll as bishop in 1786. The pope learned far more of the passions aroused in a revolutionary situation when reports reached him in 1789 of the events in France. The Church had taken its stand in support of the ancien régime in France, and its wealth made the Church a major target for the new National Assembly, which was facing an unprecedented fiscal crisis. Undercut by Gallicanism and the Enlightenment, the French clergy did little to stem the demands to solve the crisis by seizing the wealth of the Church. Emboldened by their successful assault on clerical wealth, the radicals moved on to challenge clerical privileges, the religious orders, and virtually the entire structure of the Church. In June 1790 came the Civil Constitution of the Clergy, which required that all pastors and bishops serve as elected, salaried state employees. New bishops would not ask Rome for approval but send a letter to the pope announcing their election. Pius condemned it as schismatic, which provided the rabid anticlericals their opportunity to renounce all French ties with Rome and proceed with creating the Constitutional Church.

When Austria and Prussia declared war on France in 1792, the clerics who refused to take the oath to the new constitution were deemed traitors and liable to imprisonment and execution. The radicals moved from creating a national church to de-christianizing France entirely. The papacy was identified as an enemy of the Revolution, although Pius had only a small role in the anti-French coalition. In 1796, although the worst excesses of the Revolution were over in France, Napoleon Bonaparte brought them to Italy when he defeated Austria and established republics in Milan and the Papal States. Pius was forced to sign a humiliating treaty that allowed France to intervene in Rome to protect French citizens. When in late 1796 a French general was killed in Rome in violence involving revolutionaries seeking to create a Roman republic, the French army marched into the city and ordered the pope to leave. When the elderly pope refused on the grounds that he wanted to die in Rome, the French commander replied that a man could die anywhere. The French took Pius to Florence, but when his presence there rallied the conservatives, the dying man was dragged to France, where he died on August 29, 1799. His death certificate noted the demise of "Citizen Braschi."

Pius VI was pope for over twenty-four years and six months, the longest reign yet. Had he reached "St. Peter's years" (twenty-five years), it would have given impetus to the idea that he was the final pope, which even many Catholics thought possible. Pius was not a great pope, but he did the papacy a great service by drawing up a bull in 1798 modifying the rules for a conclave under extraordinary circumstances. It waived the inaugural ceremonies, the requirement that it meet in the place where the pope died, and the time limit set for its start. Faced with the possibility that groups of cardinals scattered across Europe might elect their own popes and create schisms, Pius gave Gianfrancesco Albani, the College's dean, the authority to determine when and where to hold the election, with the proviso that it meet in a city controlled by a Catholic ruler.

There were thirty-six Italian and ten non-Italian cardinals at Pius's death. One of the six who had renounced the office asked to participate in the election, but Albani refused him. All but one of the French cardinals had fled to England, while the French had imprisoned the five cardinals who had remained in Rome. The largest group, eleven, including Albani, was in Venice.

Napoleon had given it to Austria when he seized Milan. Emperor Francis II agreed to pay the conclave expenses if it were held in Venice, and Albani was eager to have it there under Austrian protection. Albani summoned the College, and on December 1, 1799, thirty-four cardinals, including thirty Italians, met in the monastery of San Giorgio, where the physical structure of a conclave in the Vatican was reproduced as closely as possible.[2]

The College was quite evenly divided among three factions. Some cardinals wanted a pope who favored Austria, because it stood for the ancien régime that the revolutionaries had destroyed in France and much of Italy. That, however, meant accepting Austrian rule over the northern Papal States that it had occupied in 1797. The *zelanti*, conservatives who intended to rebuild Europe or at least the Catholic Church exactly as it had existed in 1789, would not accept the loss of any papal lands. Thirdly, there were the *liberali*, if that term fits men who were conservative at heart; they wanted a pope willing to make some accommodation with the Revolution. Austria had not excluded anyone, but the proximity of Venice to Vienna meant that it was easy for Austria to veto a candidate it would not accept. Franz von Herzan, who had arrived two days late, led the Austrian party; Cardinal Antonelli led the *Zelanti;* and Albani, the *liberali.* The first scrutinies saw an effort to promote Cardinal Bellisomi, whom the *zelanti* supported for his hostility to the French Revolution. He came two votes short of election. Herzan, however, had orders to elect a pope more favorable to Vienna. He prevailed on Albani to delay the next scrutiny until he received further instruction. When it came, it affirmed Austria's veto of Bellisomi. That led to a month of fruitless scrutinies in which he usually had twenty-two votes, but his reactionary attitude ensured that the *liberali* would not break ranks to make him pope. He continued to receive twenty-two votes through February.

By March 1 it was clear that someone acceptable to all parties had to be identified. Those opposed to Bellisomi agreed to select one of those voting for him, if his supporters pledged to vote for whoever was chosen. That proved to be Barnaba Chiaramonti, related to Pius VI, who had named him a cardinal. When he was unanimously elected on March 14, 1800, he became Pius VII to honor his patron. At fifty-eight years he was young compared to the median age of the previous twenty popes, but it was felt that a long reign would be a good thing for the times. A native of Cesena in the Papal States and a Benedictine monk, the new pope had raised eyebrows in 1797 when in a sermon he spoke of the need to recognize democracy as a valid form of government, which was described as "baptizing democracy." For that reason, the Austrians expressed displeasure at his election. Herzan, feeling he had already done his duty for his sovereign, had made no effort to learn if Vienna wished to exclude Chiaramonti. The emperor refused the new pope use of St. Marco Cathedral in Venice for his coronation. Pius insisted on the full display of ceremony at San Giorgio, although he had to make do with a makeshift tiara, as the French had absconded with the real one. Most of the huge crowd that attended had to stay outside; some used telescopes to view the coronation.

The emperor pressed Pius to make his residence in Vienna, but the pope felt that his place was in Rome, however disorderly and dangerous it was. By the time he reached Rome in July, Napoleon had defeated Austria in northern Italy. As the new First Consul of the French Republic, Napoleon was eager to make peace with the Catholic Church and the papacy, as a step toward securing his authority. He returned all of the Papal States to Pius and negotiated the Concordat of 1801. It restored the legal standing of the Catholic Church in France and recognized its special status as the religion of the majority of the French people. The government would appoint bishops, but the pope gained the authority to remove a bishop arbitrarily from his diocese, which was the first time outside of the Papal States and the duchy of Milan he had such power. In the ancien régime, episcopal office, once conferred, had been regarded as a property right, and it was all but impossible to force a bishop out. One consequence of the French Revolution was increased authority of the papacy over the bishops and clergy of the national churches.

The new concordat hardly created unperturbed peace between the Napoleonic regime and the papacy, but in 1803 Pius gave red hats to four French archbishops, including Napoleon's uncle. A year later Pius went to Paris to attend Napoleon's coronation as French emperor. It was by prior agreement that Napoleon himself put the imperial crown on his head. He tried to persuade Pius to stay in France by promising to return Avignon if he would reside there. Roman fears that the pope would not be allowed to return home proved groundless, as Pius left France after four months. Whatever goodwill Napoleon might have felt toward the pope dissipated, when Pius refused to support him when he went back to war in 1807. The next year French forces occupied Rome and the Papal States. In 1809 Pius was arrested for excommunicating the "robbers of the Church's patrimony" and taken first to northern Italy and then to France in 1812. Despite Napoleon's defeat in Russia, he still was able upon his return to France to browbeat Pius into signing a decree renouncing papal sovereignty over the Papal States and Rome. Napoleon intended to establish a papal residence in France; but before he could do it, his enemies were in Paris, and Pius was on his way back to Rome. When he arrived there in May 1814, thirty young men from the best Roman families took the place of the horses to pull his coach into the city.

At the Congress of Vienna that met in 1815 to reconstruct war-torn Europe, the papacy regained sovereignty over Rome and most of the Papal States. Recent events seemed to prove the necessity of papal temporal power, for without it the pope was too easily reduced to being a prisoner. This view dominated papal responses to events in Italy for the next century. The Revolution and Napoleon had swept aside much of the structure of the Church from before 1789, and much of what had been destroyed was not reconstituted. In particular those elements that had limited papal authority in the Church such as cathedral chapters and bishop-princes in the defunct Holy Roman Empire were left in the dustbin. One element from the ancien régime that did reappear was the Society of Jesus. Pius VII restored the order in August 1814 at a time when Catholic rulers were in no position to object.

The last eight years of Pius's reign were quiet compared to the first ones. Ercole Consalvi, the papal secretary of state, did marvelous work in restoring papal government in central Italy where it had been severely disrupted, without imposing a reactionary regime. He had Pius's full support, and the pope expressed his hope that Consalvi would follow him on the throne of St. Peter, yet Pius's appointments to the College undermined that goal. It was returned to full strength in 1802, when twenty-five red hats were given out, only one of which went to a non-Italian. Napoleon then pushed Pius hard on naming French cardinals, demanding that the College have at least seven. He also proposed that Catholic rulers name two-thirds of the College, leaving the pope responsible for the other third. Pius successfully withstood the pressure, but once Napoleon was gone, he named three French cardinals for a total of five. He created ninety-four cardinals in his long reign. Most proved to be more conservative than the pope, and most bore a grudge against Consalvi for streamlining the curia and reducing their clout and income. Principally by cutting the cardinals' income, Gonsalvi in 1820 achieved a surplus of 667,138 *scudi* out of an income calculated at 6,306,307 *scudi*.[3]

By 1822, when he turned eighty, Pius VII's health was failing, but he hung on until August 20, 1823, having had a reign of over twenty-three years, sixth-longest in history. He left thirteen vacancies in the College, and four cardinals did not go to the conclave, held in the Quirinal Palace because of the heat in Rome, which was unbearable in the decrepit Vatican. The practice of waiting for the non-Italians to arrive was ended; in fact the Italians showed unseemly haste in beginning serious voting while cardinals were on the road to Rome. Four cardinals entered the conclave after balloting had begun. Only Austria took an active interest in the conclave; the other states were too absorbed in dealing with several liberal revolts. Those revolts made it even more crucial for the *zelanti* to control the election. They constituted just over a majority of the voters. When the voting began on August 30, Cardinal Severoli, most respected of the *zelanti*, had twenty-one votes, enough that the Austrians had to use their veto. They objected to him, because they deemed him hostile while he was nuncio in Vienna. The Austrian leader Albani stated, "I fulfill the unpleasant duty of declaring that the Imperial Court of Vienna is unable to accept his Eminence Severoli as supreme pontiff and gives him a formal exclusion."[4] The *zelanti* were furious at this interference and expressed their outrage, but it was an element of the pre-1789 system they otherwise were eager to restore.

Consalvi, who had been de facto pope for several years, knew that he had too many enemies to be elected, but he failed to appreciate until well into the conclave how the antagonism toward him would also prevent him from being the popemaker. After Severoli's exclusion, the *zelanti* promoted Cardinal Castiglioni, admired for his fortitude while imprisoned by Napoleon.

A *zelanti* serving as scrutator recognized Consalvi's handwriting on a ballot for Castiglioni and spread the alarm. Such a blatant violation of the rules of conclave rarely was censured, nor did it rebound to the offended party's benefit. Castiglioni's support evaporated on the next ballot. The Austrian ambassador said that the conclave was full of "passion, hate, and vendetta."[5]

The *zelanti* allowed Severoli to select their next candidate, and enough cardinals voted for Annibale Della Genga out of sympathy for Severoli that he gained victory on September 28, 1823. He received exactly the required two-thirds, the smallest margin of victory in 400 years. The French were hostile to him for his abrasive efforts while nuncio in Paris to regain Avignon for the papacy, but Louis XVIII had not bothered to exclude him. That failure to recover Avignon had led Gonsalvi to remove him, which placed him among Gonsalvi's foes. Della Genga pleaded with the cardinals not to impose the burden of the papacy on him because of his poor health, and he lifted his robes to show them his ulcerated legs. When they insisted, he told them they had elected a corpse. Yet he would reign for six years.

The new Pope Leo XII, honoring Leo the Great, was born in the Papal States in 1760. Pius VII gave him a red hat in 1816. Leo was an austere, pious man who was committed to restoring the world of his youth in the Church and the Papal States, but his rule showed that he was not the typical prelate of the ancien régime. He outlawed public sale of alcohol, gambling, freemasonry, wearing of low-cut dresses, and recreation on Sundays and holy days. Draconian methods were used to eliminate crime; for example, the cardinal governing Forli ordered that brigands be executed within a day of capture. Leo restored the Inquisition and the Index of Forbidden Books. He reestablished the ghetto and special insignia for Roman Jews and required that the heads of the 300 Jewish families attend Catholic sermons on Sundays. Pope Leo was quick to avenge the wrongs done to Cardinal Della Genga. He removed Gonsalvi from power, but later came to appreciate his diplomatic talent and gave him a position through which he could use it. Leo's claims of authority over the faithful across the world led to the term ultramontane, "across the mountains," to refer to the view that the pope had absolute authority over all aspects of the lives of the Catholic people.

Leo, despite the vacancies in the College, waited a year to grant his first red hats. He named twenty-five cardinals in all; only five were not Italian. One of those for whom a red hat was sought was Abbé Louis de Rohan, from one of France's highest-ranking families. Leo refused to confer it on the grounds that he was too young, being just over thirty. The days when a youth from a princely family could become a cardinal were over. Leo's cardinals were less conservative than the rest of his work as pope might suggest. He left ten vacancies at his death on February 10, 1829. The thirty-nine electors who met in the Quirinal on February 23 consisted of thirty Italians and nine from other nations—six French, two Spaniards, and an Austrian. One cardinal turned 100 while in conclave. They were divided into three groups—liberals, moderate *zelanti,* and ultraconservatives.[6] The sixteen ultraconservatives, who placed a white cross on their cell doors, were called the "Fathers of the White Cross."

Other than expressing a desire to see a conservative but not ultramontane pope elected, the Catholic powers had only a minor role in the conclave. Two conclavists, however, were removed by force on March 15 as agents of the Austrians. The ultraconservatives were firmly united as a bloc but lacked the votes to get one of their own elected; thus they acted largely to prevent too liberal a man from becoming pope. This conclave provided a rare moment of humor in the history of conclaves when someone cast a vote for Cardinal Vidoni, a true bon vivant, who had no aspirations of becoming pope. When his name was read off the ballot, Vidoni asked in a shocked voice, "Is the Holy Ghost drunk? Who is the imbecile who is trying to make a fool of me?"[7] That was the scrutiny in which all the cardinals present received one vote apiece, probably a unique occurrence.

After a month of fruitless ballots, the popemakers agreed on Francesco Castiglioni, who had come close in the previous election. The French clergy still wanted him because of his resistance to Napoleon, and he was no longer burdened by the support of Consalvi, who had died before Leo. The arrival of four French cardinals swung the election to him. Despite telling his friends that he would not accept the tiara because of his ill health, on March 31 he was elected with thirty-two written ballots and fifteen votes by accession out of the fifty cardinals who voted. He chose the name Pius VIII, for the man who gave him his red hat in 1816 after he was freed from the prison into which Napoleon had thrown him. The new pope was almost sixty-eight years old. His election resulted in part from his poor health, since he was not expected to reign for long. This time such expectations proved correct, as Pius had only a twenty-month reign. The day of his election he wrote to his relatives ordering them not to come to Rome to seek preferment, suggesting that the fear of nepotism had not yet disappeared. His policy was more liberal than his predecessor's, reversing Leo's more draconian rules and accepting the French Revolution of 1830 that ousted Charles X and replaced him with King Louis-Philippe. Pius's reign was too short, however, to have much impact on papal policy. He died on November 30, 1830, having named six cardinals and leaving fifteen vacancies. Among his creations was Thomas Weld, who had become a priest after his wife died. After 300 years in which there rarely was an English cardinal, he began an unbroken line of Englishmen in the College. He received votes in the next conclave, the only non-Italian to do so.

Thirty-three electors assembled in the Quirinal Palace on December 14. They numbered forty-four by Christmas, when the cardinals held two scrutinies. The factions were the same as two years earlier, but the conservatives, deeply rankled by Pius's quick recognition of the liberal French king, were even more determined to win the papacy. They picked up support from several moderate *zelanti* frightened by the outbreak of revolutionary violence in France, Belgium, and Poland. Their candidates were Cardinals De Gregorio and Giustiniani, whom they thought all the Catholic powers would support. After the former failed to pass eighteen votes in three weeks, they turned to Giustiniani. On January 7 he received twenty-four votes, close enough to the required twenty-nine to make his supporters sure of victory. It came as a true shock to them, then, when the cardinal-protector of Spain announced he was excluding Giustiniani, who had served successfully as nuncio in Madrid. The reason apparently was that the cardinal had been involved in naming bishops for the new Spanish American republics, whose independence Spain had not recognized. It was suggested also that the Spanish monarchy was eager to use the veto to prevent it from lapsing. There was heated debate over whether to accept it, as many cardinals felt it was an intolerable relic of the past when kings interfered in affairs of the Church. As had happened in the past, enough cardinals agreed to accept the veto on the basis of long-standing tradition and the need to accommodate Catholic rulers.

Once it was clear that his exclusion would stand, Giustiniani spoke to his colleagues first of the favor that the Spanish king had shown him in the past and then thanked him for imposing the exclusion. (One can see why he had been a successful diplomat and was thought worthy of the papacy.)

> Of all the benefits conferred on me by His Majesty, I consider the greatest and most acceptable to me to be his having this day closed for me the access to the most sublime dignity of the Pontificate. Knowing my great weakness, I could not bring myself to foresee that I should even have to take on myself so heavy a burden. . . . Today I find myself free from my anxieties, I am restored to tranquillity, and I retain only the gratification of knowing that some of my worthy colleagues have honored me with their votes, for which I beg to offer them my eternal and sincerest gratitude.[8]

The conservatives now turned to Bartelomeo Cappellari, a member of a strict Benedictine order, but his lack of administrative experience made some reluctant to vote for him. Unlike previous monks elected pope, he had come directly from his monastery to the College and received his red hat only four years earlier. In 1795 he had penned a powerful defense of the papacy, which had won him wide praise from conservative Catholics. From the first he regularly received four or five votes and several more by accession. He now jumped up to twenty-five. While his merits were being debated and the election dragged into late January, word reached Rome of a revolt in the Papal States. Crushing the revolt would require the use of the Austrian army, since there was no longer a papal army to speak of. Cappellari came from a noble family of Lombardy, which Austria controlled, and was well regarded by Vienna. A conservative acceptable to the Austrians was clearly what was needed at that moment. He received thirty-three votes, or two more than required, on February 2, 1831.[9] He at first refused the office but was persuaded that monastic obedience required that he accept the decision of the Holy Spirit. He is the last monk to become pope and the last man not a bishop at election.

At age sixty-five the new pope was still vigorous. He took the name Gregory XVI because he admired Gregory I and Gregory VII, two of the most forceful popes in history. The name put the Catholic powers on notice that his policy would be strongly ultramontane. Gregory took a more active role than any previous pope in governing the Church around the world. He appointed bishops for the Latin American republics despite loud protests from Spain and Portugal and condemned the slave trade in 1839. Gregory's respect for established authority put him in the embarrassing position of ordering Polish Catholics in revolt against the Russian czar, who was trying to impose Orthodox Christianity on them, to submit to their proper ruler. The image of an ultrareactionary that such acts created was further strengthened by his banning the building of railroads in the Papal States, turning *chemins de fer,* the French term for railroad, into *chemins d'enfer,* "roads to hell." He also refused to allow the erection of streetlights, claiming that the people would congregate under them and plot insurrection. Yet by intervening on behalf of Catholics in such Protestant states as Britain and Prussia, he enhanced papal authority by making it routine for Catholics to look to Rome for help.

As Gregory XVI approached death in 1846, the popularity he had gained among the Romans by low taxes and public works had disappeared. His stubborn resistance to change of any sort had left a stagnant economy, an inefficient government, and the people under the heavy thumb of papal police and Austrian soldiers, who had remained in the Papal States after 1831. The papal principality that he had worked so hard to maintain unchanged was tottering under the threats of popular revolution and Italian nationalism. He was the last of the generation of churchmen formed in opposition to the French Revolution and Napoleon. He died on June 1, 1846, an old man whom time had passed by. Few other popes have been as unpopular at death. His fifteen-year reign had seen the deaths of all but twelve of the cardinals at the conclave of 1831, so Gregory had extensively remade the College. He had named seventy-eight cardinals (four more remained *in petto*), of whom sixty-six (85 percent) were Italians. Although he had taken care to give red hats to conservatives, their ideology was no longer his. Their goal was to control what change would take place, not prevent it.

Gregory had created a thoroughly Italian College. Only eight of the sixty-two cardinals at his death were not Italian, and the only non-Italian who was in Rome to vote for his successor was Cardinal Acton. Whether he properly can be called an Englishman is debatable, since he was born in Naples, spent a decade in England for his education, and returned to Italy for the rest of his life. Gregory had known few prelates before becoming pope, and most of those he came to know after were members of the curia. Thus, the fifty cardinals who participated in the election overwhelmingly were curialists. They were probably the most uniform set of electors since 1059. Absent any international presence, the College was determined to speed the election in order to prevent interference by the Catholic powers and to avoid a drawn-out conclave. French King Louis-Philippe regarded it as improper for a ruler to interfere in the papal election, and hardly expressed an opinion on the candidates, to say nothing of exercising a veto. Austria was eager to have its voice heard, but Cardinal Gaisrück arrived too late from Vienna to participate.

Luigi Lambruschini, Gregory's secretary of state, who governed for him in his last years, was the candidate of the conservatives; but secretaries of state rarely have been elected, as they are too closely identified with the dead pope. The desire for some change has almost always been a strong feeling in conclaves. The *liberali* supported Cardinal Gizzi, the favorite of the Romans but too liberal for the College. What most cardinals wanted was a man ready to make changes in the Papal States to calm the threat of rebellion, yet conservative in doctrine. When the first scrutiny took place in the Quirinal on June 15, 1846, Lambruschini received fifteen votes and Giovanni Mastai-Ferretti, thirteen. Young at age fifty-four and relatively unknown, Mastai appealed to those who wanted moderate change, because he was a devout man loved by the people in his diocese of Imola in the Papal States. Although he had not been part of the curia, his backers persuaded enough cardinals that he was the right man for the papacy for that time. Rumor had it that Gaisrück was bringing a veto against Mastai, and his supporters redoubled their efforts to get him elected before the Austrian arrived. In the second scrutiny, the numbers from the first were reversed, and at the fourth scrutiny Mastai received twenty-seven written ballots and seven by accession, or two more than required. Mastai was a scrutator for the final ballot and had the rare distinction of announcing his own election as pope.[10]

The new pope, born in 1792 into a noble family from the Papal States with a tradition of service to the papacy, had intended to pursue a military career, but his health was too fragile and he turned to clerical life. After becoming a priest in 1819, he was included in a papal delegation to Chile and Peru, becoming the first future pope to have set foot outside of Europe since Gregory X had been elected in 1271 while in Palestine. Although the delegation failed in its task of regularizing relations with the new states, Pius developed a deep interest in the missions, which led to his designation as "the missionary pope." After his return he was named archbishop of Imola in 1832. Gregory XVI gave him a red hat in 1840 for his work in keeping Imola quiet while rebellion erupted around it.

In the evening of June 16, word spread that the election was about to be concluded, so the next morning a huge crowd gathered for word of the new pope. Rumor had identified a local favorite as the new pope, and a mob sacked his house. The practice had not disappeared, although there is no mention of it for the preceding century. The new pope ended it by establishing the practice of giving a gift to the curia staff of a month's pay at a pope's death and again at the election of the new one. With a round of cannon fire from the Castel Sant'Angelo, the new Pius IX was presented, which name Mastai had taken to honor Pius VII, his first patron. The crowd was reticent in its cheering, as the Romans knew little about him. Soon he became wildly popular, for he seemed to be everything they could want in a pope. Generous, affable, and more liberal than most cardinals expected, he gained the adoration of the crowd when he issued a general amnesty to those who had been imprisoned or exiled under Gregory for sedition. Expectations were raised higher when Pio Nono, as he is usually called, named Gizzi as his secretary of state. Reform in the administration of the Papal States and a general atmosphere of change, for example, building railroads, seemed to confirm that this was a liberal pope.

How long Pio Nono would have continued reforming the Papal States and the Church were it not for the events of 1848 is impossible to say. He was more conservative than he appeared until November 15, 1848, when radical Italian nationalists in Rome murdered his newly created prime minister, a layman. Nine days later the pope was fleeing to Naples. He expected

Austria to restore him to power, but it was France that did so, as Napoleon III, just established in power, was eager to be seen as the defender of the Church and to undercut Austrian power in Italy. In April 1850, Pio Nono returned to power in Rome, but only as long the French troops remained, while the new kingdom of Italy had annexed most of the Papal States. The pope believed that he had given Italian nationalists and liberals the opportunity to show that they could act responsibly in creating an Italian state, and that they had failed dismally. He therefore was free to oppose both. His new political conservatism prompted the same in the Church, as he no longer was concerned about what Catholic liberals thought of him in respect to church affairs. His appointments of cardinals reflected his new attitude.

The promulgation in 1854 of the doctrine of the Immaculate Conception of Mary reflected this new attitude, since its opponents were mostly liberal Catholics. They believed that the pope did not have the authority unilaterally to declare doctrine without consulting the bishops, and that the doctrine undermined efforts to reconcile the Protestant churches with Rome. The objection that the pope lacked authority to establish doctrine rankled Pio Nono, and he set about to remedy the lack of a clearly defined statement of such a right by convoking a council for 1869 in the Vatican Palace to define papal infallibility: A pope speaking *ex cathedra* (from the chair of St. Peter) could not err in defining doctrine or morals. Catholic liberals but also some conservatives objected on the grounds that there was no support for the doctrine in Scripture or tradition. Pius's intransigence on the issue led some of them to become schismatic, establishing "Old Catholic" churches in northern Europe. His success in winning the approval of most of the bishops at Vatican I demonstrated that Catholics largely accepted the doctrine.

Vatican I largely coincided with two other momentous events. On September 20, 1870, with the Franco-Prussian War making French aid to the pope impossible, Italian troops entered Rome to claim it as the capital of a unified Italy. Pius refused to accept the guarantees of safety, use of the Vatican Palace, and financial aid that the Italian government offered him, and confined himself to the Vatican. He, and most Catholics, referred to himself as the "Prisoner of the Vatican," although the Italian government would have been pleased if he had come out and carried on his duties in Rome. Being closed up in the Vatican pained Pius, since he was a gregarious man who loved mixing with the people. Six months later he celebrated the twenty-fifth anniversary of his election. A pope at last had seen "St. Peter's years." The statement in the coronation ceremony that the new pope "will not see St. Peter's years" was dropped. Pius was said to have snapped, "That is no article of faith!" when it was said to him at his coronation.[11]

Not a few Catholics wondered whether the loss of papal temporal power and the breaking of a barrier thought impenetrable were signals for the end of the papacy. Of course it did not, but the papacy after 1870 was a very different institution. Pius and most Catholics objected to the loss of Rome, because it seemed clear to them that the pope had to have sovereign rule of a state in order to maintain his freedom of action in governing the Church. In fact the pope gained a great deal more freedom by losing his temporal realm. So many decisions of the past had been based on the popes' concern for losing the Papal States. Now the pope was free do what was best for the Church and his authority in it, not for his temporal power. By losing his absolute power over a temporal realm, the pope became absolute in the Church. One can argue that Count Cavour and King Victor Emmanual did more good for the papacy than Pius himself did.

After 1870 improvements in communications, the independence of the Latin American republics, and the rapid rise of Catholic populations in Britain, the United States, Canada, and Australia all required that the pope look at the world beyond continental Europe as more than missionary territory. Pius's appointments to the College began to reflect that, as he gave a red hat in 1875 to John McCloskey of New York, the first for the United States, and he named the first cardinals from an Irish see and the Ukrainian Uniate Church. Because of his long reign, Pius named 123 cardinals. They included fifty-two non-Italians, the highest proportion since the mid-sixteenth century.

Early in 1878 Pio Nono turned eighty-six, still vigorous for his age, but his health suddenly deteriorated, and he died on February 7 after a reign of thirty-one years and eight months, the longest ever by an impressive margin of six years. Of the sixty-four cardinals then in the College, only four had participated in the conclave that elected him, and one of them was too ill to attend. Besides the limited institutional memory of the workings of a conclave, there was another problem: The king of Italy had taken over the Quirinal palace for his residence. Before the cardinals could decide in which building the conclave would take place, they had to decide whether it would even be held in Rome. Some who were hostile to the new Italy demanded that it be moved out of Italy entirely. Cardinal Manning wanted the College to ask the British government to let it meet in Malta. Others raised the threat of a schism, arguing that if the conclave were held away from Italy, there would be at least a few cardinals willing to elect a pope in Rome, recreating the situation of 1378. The Italian government declared publicly that it would not interfere in any way in the conclave, while it privately threatened the cardinals that if they left Rome, they and their pope would never be allowed to return.

The thirty-eight cardinals present in Rome met two days after Pius's death and voted to require that the new pope swear to defend his right to the Papal States "even to the shedding of blood." They also agreed, with only five dissenters, to hold the conclave outside of Italy. The next day, after having time to reflect on the ramifications of that decision, they voted to meet in the Vatican. The palace was even more decrepit than it had been in 1820 when the decision had been made to meet in the more comfortable Quirinal. Materials for the cells were in the Quirinal, and the College refused to ask the Italian king for permission to move them. The building of the cells had to be done with the limited supplies in the Vatican; over 500 men worked on making it ready. All the cardinals except for one, a monk, later complained about the discomfort of their cells. On the first day of the conclave, Vincenzo Pecci, who had became the *camerlengo* in 1877, thereby responsible for organizing the conclave, ordered changes in the design of the voting chamber, which required workmen to labor through the night. Since Pecci's cell was closest to the work, it kept him awake until he ordered it halted.[12]

Reporters were allowed into the conclave space before it began, and one commented that it was like a great medieval fortress, with Swiss Guards holding halberds at every turn.[13] One innovation that resulted from the move back to the Vatican was the use of a common kitchen for meals for the conclave. No longer would the cardinals have to arrange for their meals to be brought in from the outside. The change reduced the opportunity for communicating the doings of the conclave to the outside world. A German cardinal had an aversion to pasta and insisted on having his own cook. The others complained that if they had to eat the common food, he should, too. The conclavists and the staff ate together at their own table, but the clerics objected to eating with laymen, and they were given a separate table.

Despite all of the difficulties, the conclave began on February 19, only two days late. The short delay allowed sixty cardinals to reach the Vatican. The Mass of the Holy Spirit, traditionally offered in St. Peter's, was said in the Vatican instead. Nor did the dean say it, as was usual, for Cardinal Amat was so ill that he was carried into his cell and never left it for the duration of the conclave. The College no longer included any members of the great Roman families of the past, although there was a Cardinal Bonaparte. One Italian and three non-Italians were absent; two, including McCloskey, rushed to Rome but did not make it in time. Twenty-two non-Italians voted, among whom the six Austrian cardinals were the largest bloc. For the first time since the late sixteenth century, Italians constituted less than two-thirds of the electors. It gave rise to discussion in the *prattiche* whether a non-Italian ought to be elected, to demonstrate the cosmopolitan nature of the Church and to spite the Italian government. The Italians, informed of the talk but pledged not to interfere, asked France and Austria to use their exclusion to prevent such an outcome. Austria and Spain insisted on their right of veto, but the French republic did not regard it as proper for a secular state to use. The cardinals debated whether the exclusion still existed, and the majority conceded that it did.

Although Pius IX had appointed nearly all the electors, there was a consensus that a different type of pope should be chosen; that view also reflected deep antipathy to Cardinal Antonelli, Pius's right-hand man, who had died the year before. The loss of political power and the likelihood that the new pope would probably also have to remain "the prisoner of the Vatican" reduced the number of those who sought the tiara and thus the politicking found in previous conclaves. There was little question even before the conclave began that the best man for the job was Vincenzo Pecci, who had been made a cardinal in 1853 and became the *camerlengo* in 1877, thereby responsible for organizing the conclave. He was nearly sixty-eight, so the younger cardinals felt secure that there would not be a repeat of Pius's incredibly long reign. Pecci was seen as a moderate conservative who would accept change when the need for it was clear. Still, there was opposition to him. Cardinal Bilio, Pius's confidant in his last years, rallied the most conservative cardinals behind Cardinal Martinelli, a saintly man with no experience in politics. His naïveté was so obvious that Pecci's supporters declared he would be another Celestine V.

When the first ballot was counted, there were two defective ballots, and it was declared invalid; but the cardinals insisted on a count to see in what direction the balloting was going. Pecci had nineteen votes. His count rose to twenty-six in the second, and in the third he reached thirty-six, or three less than needed. After the ballots were burned, his supporters rushed to acclaim him as pope, and enough others did as well to give him the election. The sources disagree on whether all who had not voted for him acclaimed him. Pecci had the previous day expressed deep reluctance to take on the burden of the papacy, but now he quickly agreed, announcing that he would be called Leo XIII to honor Leo XII. His reason was never made clear, since he was only nineteen in 1829, when Leo died, and Leo gave no patronage to the new pope's family, from Carpineto in the Papal States. Since there already had been the smoke signaling an unsuccessful ballot just prior to noon, St. Peter's square was nearly empty when the dean stepped out on the balcony to announce "Habemus papam!" Leo, who required rest after the stress of the election, did not give his first blessing until 4:30. He chose to give his first blessing *urbi et orbi* inside St. Peter's. It was a sign that he had no intention of changing the antagonistic relationship between papacy and Italy, made clear by his refusal to inform the king of his election. Leo confirmed Pius's ban on Catholics' voting in Italian national elections, but he allowed it on the local level.

It seldom has happened in papal elections, but the cardinals got what they expected in Leo XIII—a pope adamant about regaining temporal power but only one step behind the world rather than the two or three steps of most popes since 1820. Despite having had only a brief posting as nuncio to Belgium, Leo proved to be a skilled diplomat. He negotiated a reduction in tensions with Germany, France, and Portugal, where the governments were opposed to clerical influence on public affairs. There were, however, no negotiations with Italy, for the pope would accept nothing less than a complete surrender on the issue of Rome. Being a solitary man, confinement in the Vatican bothered him less than it had Pius, and there he remained for twenty-five years.

While refusing to back off Pius's attacks on modernism, Leo reduced the hostile tone. In his *Rerum Novarem* (1891), regarded by many as the greatest papal encyclical, he made clear his belief that the Church needed to demonstrate concern about the welfare of the lower classes of society. He recognized the rights of the laborers to organize for that purpose. Catholic political parties and labor unions began to appear in such lands as Belgium and Germany, which provided an alternative to atheistic and anticlerical socialist movements, especially Marxism. In this and several other encyclicals, Leo acknowledged the legitimacy of democracy as a form of government for secular states, but most certainly not for the Church. He was more of an absolute monarch in the Church than any of his medieval predecessors had been.

CHAPTER X

Conclaves in the Twentieth Century

The pattern of papal elections after 1900 reflected the new stateless papacy imposed on Pius IX and Leo XIII. Both by losing control of the Papal States and legitimizing democracy in secular politics, the papacy moved away from the fusion of ecclesiastical and secular authority that had been a curse for the Church for a millennium. The pope's official authority became strictly moral and spiritual, yet the popes after Leo found that papal prestige and influence reached far higher levels than had been true since the time of Innocent III. Stalin's crack, "How many divisions does the pope have?" was relevant for the nineteenth century, but his successors would find that the papacy of the late twentieth century wielded enormous authority in politics by exercising a strictly spiritual office.

In March 1903, Leo XIII celebrated the twenty-fifth anniversary of his election. After 1,800 years of no pope seeing "Peter's Years," two popes in succession had passed that mark. Unlike Pius IX, Leo did not live long after his jubilee, dying on July 20. Leo's long reign allowed him to remake the College by naming 147 cardinals. Nearly the same proportion was Italian— 58 percent—as for his predecessor. Natives of the old Papal States constituted more than a quarter of the College; nine came from Rome. Leo's notable firsts in the College included cardinals from Australia, Canada, and the Armenian Rite. Only one of his predecessor's creations remained when the conclave began on July 30. All but two of sixty-two cardinals attended. The Cardinal of Sydney was still at sea, and the Cardinal of Palermo was too ill. James Gibbons of Baltimore, the sole American cardinal, happened to be in Rome and became the first American to vote in a papal election.[1]

The development of express trains since the last conclave allowed European cardinals to reach Rome in time not only for the first scrutiny but also for the *prattiche,* which previously had been largely restricted to the Italians. Within the College the major parties were the conservatives and moderates. There were two or three cardinals who could be called liberals in that they were prepared to accept change more willingly than their confreres. Eight cardinals were regarded as *papabili.* Mariano Rampolla, Leo's secretary of state, had support among most non-Italians, who appreciated his talent as a diplomat. The Austrians, however, opposed him for being too pro-French, and conservatives felt he had been too favorable to the anticlerical Third French Republic. There was also the usual problem for a secretary of state— that of being closely identified with the dead pope's policy, much the greater when the pope had reigned for twenty-five years and the secretary of state had served for fifteen. Rampolla was too much like his master—an austere intellectual who had little experience in the care of souls. Since 1800 the choice of popes had alternated between a pastor and a diplomat, and it was time again for a pastor.

At the first scrutiny Rampolla received twenty-four votes, a strong showing for a first ballot. The number of votes needed for election was forty-two. An elector wrote "no one" on his ballot, which touched off a lengthy debate about whether it invalidated the entire scrutiny; it was decided that it did not. Rampolla reached twenty-nine on the second ballot, but as the cardinals assembled for the third scrutiny Cardinal Puzyna of Cracow, which was then under Austrian rule, rose to pronounce that Austria excluded him. Puzyna had tried to get both the dean and the conclave secretary to accept his statement of exclusion, but neither would. He was obliged to read it himself to the cardinals. He had been given the task, because he opposed Rampolla's election. It is possible that Italy asked Austria to do it. The dean, although opposed to Rampolla, declared that the veto would have no bearing on the election. While disclaiming any ambition to become pope, Rampolla objected sharply to this intrusion of a secular state into the affairs of the Church, but the College as a whole showed little outrage. Bulls by prior popes banning coercion in the conclave had not explicitly mentioned exclusion, and, therefore, it was deemed legal. Rampolla's supporters hoped the veto would not be effective, and in the third scrutiny Rampolla gained another vote, giving him thirty, but it was not a meaningful gain at that point in a conclave. Giuseppe Sarto of Venice, who had started with five votes, now had twenty-four. At the fifth scrutiny Sarto was up to thirty-five, while Rampolla slipped to sixteen.

Sarto's cell in the Vatican was located next to a French cardinal's, who had asked him at the start of the conclave if he spoke French. When Sarto replied that he did not, the Frenchman declared that he could never be elected, since speaking the language of diplomacy was essential in a pope. Yet now the French agreed to vote for Sarto, pushing him past the needed majority. In the first ballot on August 5 he received fifty votes, eight more than necessary for election; Rampolla finished with ten. As Sarto had gained votes, he had protested with tears that he was unworthy to become pope; but by the last scrutiny he had made peace with the idea. He announced that he would be called Pius X, to honor the earlier Popes Pius who had suffered so much for the Church. At 11:15 AM, which had been the usual time the previous days for the appearance of the smoke from the burning of the ballots, there was no smoke. Yet many outside knew by then of the election's completion, because the chief steward did not order food for that day's supper, indicating that the conclave would be over. Conclavists were signaling from the windows that Sarto had been elected by imitating the sewing motions of a tailor, *sarto* in Italian. At 11:40 the news of the new pope was given to the crowd from the outside balcony of St. Peter's, but his first blessing was done within the basilica. It indicated that Pius had no intention of changing in any respect the adversarial relationship with the Italian state, a point emphasized by his refusal to inform the king of his election. He remained "the prisoner of the Vatican" to his death.

The new pope was born in 1835 in a village near Padua. One of eight children of a mail carrier and a seamstress, he was ordained as a priest at the young age of twenty-three. In 1884 he became bishop of Mantua, where he earned a reputation as an effective prelate in the midst of chaos. His appointment in 1893 as patriarch of Venice came only after three prior choices refused the office because of the difficult political situation there. That office always carried with it a red hat. He spent little time in Rome during his ten years at Venice, where he was known as a pastor, not a politician. His experience in the curia was the least of the Italian cardinals, but it was precisely because the College saw the need for a pastor in Saint Peter's chair that it turned to him. Lacking diplomatic experience, Pius named Rafael Merry del Val as secretary of state, whose views on the need for the Holy See to keep its distance from secular governments agreed with his own.

The Austrian exclusion of Rampolla offended Pius because it was an act of blatant secular interference, although it insured that he would become pope. In early 1904 he wrote a bull excommunicating anyone attempting to use it at a future conclave, but the bull was not made public. At least two reasons persuaded him to keep it unpublished: The Austrians would be highly annoyed if there was a public declaration that they had acted so sinfully; and Pius's own election would be brought into question if exclusion was so evil that excommunication was the proper punishment. In 1908 a Catholic newspaper revealed the existence of the bull, and it was published the next year. By then it created little stir. The exclusion probably would have disappeared, anyway.

The extensive information on the conclave that appeared in the newspapers also scandalized Pius. The results of the scrutinies and news of the Austrian exclusion appeared in Roman papers within twenty-four hours. Pius was a firm believer in papal authority, which had dictated that deliberations and votes be kept strictly secret. He therefore issued a new decree on the conclave procedures that contained little that was new but reinforced with the strongest language the demand for complete and permanent secrecy. The decree's major point was automatic excommunication for anyone who revealed what had happened in the conclave even well afterward. Thus, at a time when political interference—the original reason for secrecy—was ending, enforcement of secrecy was made all the stronger, and it has been well obeyed since 1903. Now journalists were the ones who were seeking to find out what was going on, and the reaction of the popes has been to keep them at bay by emphasizing the obligation of secrecy. One result has been that information on vote counts is less extensive for the elections after 1903 than for those of 500 years earlier.

Pius X was perhaps the most pastoral pope of the past millennium. His concern was primarily the faith of ordinary Catholics, which he believed could be greatly strengthened by frequent reception of the Eucharist. He lowered the age for children to receive their first Communion to seven, and strongly encouraged adults to receive the sacrament weekly if not daily. His pastoral concerns naturally extended to doctrinal purity among the faithful. He sought to ensure it by strengthening the Congregation of the Doctrine of the Faith, the Inquisition's new name, and defining as heresy the tenets of "Modernism," largely the belief in democracy and the application of scientific criticism to the study of religion. In order to make sure that the faithful were being lead by true pastors, he drew up an oath denouncing Modernist heresies and required all clergymen to swear to it. He encouraged the creation of secret societies of clerics and pious laymen opposed to Modernism, who called themselves "Integral Catholics," to inform on those suspected of Modernism. The result was a rigid orthodoxy that ruined the careers of many Catholic theologians who were seeking to use the new scholarship in biblical study.

Pius's doctrinal conservatism spilled over into his appointments of cardinals. The most capable and among the most conservative was Merry del Val, who was given a red hat at age thirty-seven, the youngest cardinal of the twentieth century. Pius named fifty in all, including the first Latin American, a prelate from Brazil, but a majority were Italian. Although he wished to appoint "Integrists," he also was interested in having cardinals from a broader range of countries. The two goals proved to be contradictory in the next conclave, since most of the non-Italians opposed the Integrists, largely based in the curia, because of the bad relations between it and the national churches. When Pius died on August 20, 1914, he took the names of two cardinals to the grave. He had ordered that his body not be embalmed, a practice followed for the next four popes. The long time before they were buried resulted in several unpleasant situations. Within a decade of Pius's death, the process for his canonization began, which was completed in 1954 when he was declared a saint, the first pope so honored since Pius V.

Pius X had the pain of seeing the outbreak of World War I and the good fortune not to know how long and destructive it would be. It was clear that a diplomat would be elected because of the war. It was involving more countries almost daily, although Italy was still neutral as of August 31, when the conclave began. It was the first one held during a Europe-wide war since 1758, and no one knew how the cardinals from the warring nations would respond to being locked together in such close quarters. To their credit, there is little hint of unpleasantness, and the cardinals seem to have behaved better than their distant predecessors.

There were sixty-five cardinals in the College, but two were ill and did not attend, and six failed to reach Rome in time because of the distances they had to cover. Efforts to delay the conclave to allow them to reach Rome in time were derailed by the argument for quick election, so the cardinals of the warring nations could hurry home. It was also argued that it was necessary for the College to present an image of unity by agreeing quickly on a new pope. Two American cardinals, Gibbons and O'Connell, booked passage on a liner that sailed the same day they heard of Pius's death. (When they reached Naples, they rented a car and sped toward Rome, but ringing church bells in a village they were passing through alerted them that the conclave was over.[2]) Farley of New York was already in Europe and joined the contingent of twenty-five non-Italians at this election. Construction of the cells was complicated with the discovery that the furnishings from 1903 had been given to the poor, but the problem was solved in time by renting what was needed from a hotel.

The conclave of 1914 demonstrated the effectiveness of Pius X's edict obliging secrecy. Information is limited, and the sources give varying numbers of votes for the candidates. The war also required the cardinals to be more careful about leaking information about their deliberations and voting. Although Merry del Val was regarded as too young to become pope, he was the most powerful person in the conclave. Yet he could not push through the election of his choices because of the usual objections to continuing the power of the dead pope's secretary of state. He and the other Integrists supported Basilio Pompilj, while the nonintegrists had two major candidates in Pietro Maffi and Giacomo Della Chiesa. Both had twelve votes in the first scrutiny, while Pompilj had nine, and Merry del Val, seven. In the second both Maffi and Della Chiesa rose to sixteen each, while Pompilj only reached ten. Merry del Val then promoted Domenico Serafini, who rose from two votes in the first ballot to twenty-four by the eighth. Those opposed to Merry del Val began to focus on Della Chiesa, who had waited seven years for his red hat after becoming archbishop of Bologna in 1907,which was a see that traditionally conferred the cardinal's hat on a new archbishop. Merry del Val had opposed giving it to Della Chiesa, because he was Rampolla's protégé. It came only after Rampolla died in 1913. Because of his thirty years' service in the papal diplomatic office and the belief that he was hostile to Merry del Val, he emerged as a strong candidate. The German and Austrian cardinals voted for him en bloc.

Although he had thirty-two votes in the eighth scrutiny, the conclave went to ten scrutinies, because Della Chiesa's opponents argued that he was "a good bureaucrat but a mediocre man," whose election would be an affront to the late pope.[3] Faced with a choice between him and Serafini, who was only fifty-one, several Integrists voted for Della Chiesa, making him pope on the morning of September 3. The strongest evidence of his vote total came from the pope himself. Some close friends had remained in St. Peter's square long after the expected time for the morning ballots, because there had not yet been any smoke. Speaking later to the new pope, they mentioned that they had waited despite a general belief that no one had been elected. He replied, "Oh, you waited, did you? They had to open my ballot to see for whom I had voted."[4] The only reason why it would have been checked was if he had exactly a two-thirds majority—thirty-eight votes for this election—because a cardinal was not allowed to vote for himself. This means that Della Chiesa was elected with the smallest possible number. It was apparently Cardinal de Lai, Merry del Val's ally, who asked for the check. It was a stinging insult to the new pope, and if he blamed Merry del Val, whether or not he had any role, it would explain why the cardinal was completely eclipsed after 1914.

Accepting the tiara without any hesitation, either real or feigned, Della Chiesa declared that he would be called Benedict XV, without explaining his choice, the first name other than Pius, Gregory, or Leo since 1774. He usually is said to have honored Benedict XIV, also an archbishop of Bologna. Benedict came from a noble family of Genoa. His frail body, limp, and sloped shoulder were attributed to the difficulty of his birth in 1854. Although he wanted to become a priest, his father insisted that he study civil law, and he received a law degree before joining the clergy. In 1882 he joined Rampolla in Spain. When Leo XIII named Rampolla secretary of state, Della Chiesa was promoted to undersecretary. His close affiliation with Rampolla led to his dismissal when Pius X was elected. Soon he was exiled to Bologna. His red hat came barely two months before the conclave. Cardinal Gibbons, on the news of his election, asked, "Who's he?"

Benedict's talent as a diplomat was the deciding factor in his election, since the war made it sorely needed. His careful neutrality was made clear in that the French called him the "German pope," and the Germans, the "French pope." When Italy joined the Allies in early 1915, papal neutrality became all the more difficult to maintain. A German offer to help the pope recover the Papal States did little to endear him with the Allies. In August 1917 Benedict sent the warring nations a seven-point peace plan that would have returned Europe to the status quo of 1914, while also calling for freedom of the seas, a reduction in armaments, and a forum for international arbitration. Germany showed some interest in it, but the Allies dismissed it out of hand. Its similarities to Woodrow Wilson's Fourteen Points are intriguing, but the American president never admitted to having been influenced by the pope. The stern Calvinist Wilson probably would have never admitted it to himself.

Because of objections from Italy and the perception among the Allies that the pope was pro-German, he was not represented at the negotiations that resulted in the Peace of Versailles, which he criticized as too vengeful. Italy feared that the pope would use the negotiations to recover Rome, which issue continued to sour relations between papacy and the Italian state. Benedict made an effort to reduce tensions by permitting Italians to vote in national elections and allowing non-Italian Catholic princes to visit the Italian king without being denied an audience with the pope. Under Pius X the relationship with France had been so strained that the French withdrew their ambassador from the Vatican. Benedict's most effective act in restoring ties to Paris was Joan of Arc's canonization in 1920. She had been promoted as a symbol of French patriotism during the war, and the French saw her sainthood as belated papal recognition of the justice of their cause. In addition, Benedict reduced the influence of the Integrists in the Church by ending the system of secret informers set up under Pius.

A new code of canon law, which Pius X had ordered undertaken, was finished under Benedict. It codified the system of the conclave as used in 1914 and the powers of the cardinals, defined as the sole papal electors. The prerequisites for becoming a cardinal included for the first time ordination to the priesthood. The College was kept at seventy members. Benedict had less opportunity to remake the College, naming thirty-two cardinals, and there were no notable "firsts" among them. Their nationality breakdown was about the same as for Pius—just over half Italian. The opportunity to participate in a conclave came more quickly than anyone expected. Benedict, who missed hardly a day at work in seven and a half years as pope, caught a cold in early January 1922. It turned into pneumonia, and he died on January 22.

The cost of the conclave became an issue because of the dire straits of papal finances. Even before it began, its expenses were placed at 2 million *lire*, including 800,000 for the physical work in the Vatican, despite such economies as taking furniture from nearby hospitals. Nuns were recruited as cooks, because they asked for less pay than monks. The poor quality of their cooking was still a complaint among cardinals in 1978. Telephones were put in the cells, but only so that the cardinals could talk with each other; there was no access to an outside line. Italy, France, and Spain declared that the veto was a thing of the past, while the new republic of Austria was not represented in the conclave. Cardinal Bourne of England was noted as a *papabile*, which led Irish Cardinal Logue to declare, "It was an English-speaking Pontiff who gave Ireland to England in the thirteenth century; therefore, we do not favor any but an Italian for the throne of St. Peter."[5]

On February 2, fifty-three cardinals were locked in the Vatican, but only after a photographer was discovered hidden inside and expelled. The conclave officials also foiled a reporter's plan to take the place of a waiter inside the Vatican. The four cardinals from the Americas did not reach Rome in time. O'Connell, one of those who missed voting in 1914, had devised at a moment's notice an elaborate plan for getting to the conclave. A sea liner was held at New York until he could get there from Boston by airplane, making him probably the first cardinal ever to fly. He then met a faster Italian liner at Brest, France, which whisked him to Naples, where a special train to Rome was waiting for him. As he ran from the train at Rome, he was informed that the election was over. His rage was monumental, which he expressed to the new pope. It had the desired effect of persuading the new pope to change the waiting period from ten days for the first scrutiny to fifteen days, with the option of extending it to twenty days if the College so chose. O'Connell voted in the next conclave.[6]

Had O'Connell reached the conclave in time, his would have been another vote for Merry del Val, although it is unlikely that it would have changed the outcome. Although Merry del Val was now the proper age, he never passed seventeen votes, while Cardinal Gasparri, the *camerlengo,* peaked at twenty-four. O'Connell later accused Gasparri of rushing the balloting to prevent him from voting for Merry del Val, but Gasparri had in fact proposed to delay the conclave, which was rejected with much less justification than in 1914. Years later the beatification process for Merry del Val was halted because of evidence that he had broken conclave rules so severely he had merited excommunication. Since there is no hint that he violated secrecy, the only other possibility is that he solicited votes.[7] With neither man capable of getting elected, the cardinals turned to others. La Fontaine of Venice and Ratti of Milan emerged as the leaders. La Fontaine, supported by the conservatives, reached twenty-three votes before falling back. Ratti had nine votes in the first scrutiny but had dropped to four in the seventh, as Gasparri had most of the progressives' votes. By the eleventh, with Gasparri now actively but covertly promoting him, he was up to twenty-four. Three ballots later, at the second scrutiny of February 6, Achille Ratti became pope with forty-two votes, six over the required number. He announced that since he was born in 1857 under Pius IX and became a priest under Pius X, and because the word meant peace, he would be Pius XI, although he would continue Benedict's policies.

More than most popes, Pius XI was the protégé of his predecessor; preconclave reports mentioned him among the *papabili* who would continue Benedict's policies. He came from a middle-class family near Como. Pius X appointed him head of the Vatican Library in 1914. That position required him to work closely with Benedict, and they became friends. In February 1918 the pope shocked Ritti by naming him apostolic visitor to Poland despite no diplomatic experience. Poland was in the throes of organizing a government as the three empires that had occupied it since 1795 collapsed after World War I. He was in Warsaw when the Red Army reached its outskirts in 1920. It convinced him of the dangers of Communism. Caught up in border disputes between Poland and Germany and accused by both of being too partial to the other, Ratti was recalled in 1921 and named archbishop of Milan, a see that conferred a red hat on its incumbent. A half-year later he left Milan to attend the conclave, saying as he departed, "Off we go to prison."[8]

Pius is best known for making peace with the Italian state, which he signaled by giving his blessing *urbi et orbi* from the outside balcony of St. Peter's for the first time since the loss of Rome in 1870. The head of the Italian government was Benito Mussolini, whose Fascist policies troubled the pope. It took him four years to begin negotiations with Mussolini and another three years before the Lateran Treaty was signed in February 1929. It recognized papal sovereignty over the 109 acres of Vatican City, along with the Lateran and Castel Gondolfo and the right to have a post office and radio station. It gave the Church extensive rights over Italian education and marriages and a vast sum as an indemnity for the Papal States. Besides recognizing Italian sovereignty over the former Papal States and Rome, Pius agreed to abandon his support for a Catholic political party, which was the principal source of opposition to the Fascist party.

Much of the non-Italian hierarchy felt that Pius had conceded too much in the treaty and criticized him for dealing with the Fascists. Whether or not an earlier pope could have made the same deal with a prior government, it has been of enormous benefit for the papacy. It gave it the independence from secular control that the popes had always claimed they needed, while removing nearly all of the concerns and problems of ruling a large state. The concordat signed four years later with Germany was harder to defend, since it produced little benefit for the Church at the cost again of abandoning a Catholic party that opposed Hitler. Pius was certainly no democrat, but by 1936 he recognized the threat to religion from the regimes in Germany,

Italy, and the Soviet Union and became outspoken in opposition to them. He wrote an encyclical to the German church in 1937 against Nazi racial policy, which failed to achieve anything. In 1939 he wrote a powerful denunciation of the Nazi treatment of non-Germans, but died before delivering it, and its contents remained unknown for almost three decades.

Pius XI gave red hats to seventy-six men; just over half were Italians. His creations included the first Asian cardinal in over a millennium, in the person of the patriarch of the Syrian Rite Catholics, and the first from Argentina. At his death in 1939 there were five cardinals from the United States. Pius reappointed Gasparri as secretary of state, making him the only man to serve two popes in that office. When Gasparri died in 1930, his protégé Eugenio Pacelli received the office. Pius worked to win him the tiara. He sent Pacelli on numerous missions around the world, partly because he spoke many languages fluently, but also, as Pius explicitly stated, so that the non-Italian cardinals could get to know and appreciate him. Pius rarely called the noncurial cardinals to Rome, so the only member of the curia that most knew was Pacelli. Of course, that could have resulted in antipathy toward the man, but he was a talented diplomat and an impressive churchman, so when the cardinals gathered in Rome after Pius's death on February 10, 1939, he was the overwhelming favorite. A week before Pius died, he said to a secretary that Pacelli "will make a fine pope."[9]

The new rule on delaying the conclave put the beginning of the conclave on March 2. This time O'Connell, who again came by sea, reached Rome in time to participate, along with three other Americans, a Canadian, and two Latin Americans. In all there were sixty-two cardinals. This conclave saw the end of the tradition of marking with purple or green the thrones of the cardinals according to who had appointed them; all now had purple. Before the conclave began, it was estimated that it would cost $100,000; its brevity suggests that it probably did not surpass that.

Most of the Italian cardinals with their own sees favored Cardinal Dalla Costa of Florence, because he was viewed as unworldly and nonpolitical and would not upset the delicate relationship between church and state in Italy. The Fascists and Nazis regarded Pacelli as responsible for Pius's censures of them. But the time when the Italians would determine a papal election by themselves had disappeared. The curialists and non-Italian cardinals, even the Germans, supported Pacelli, giving him thirty-six votes in the first scrutiny. The Italians began swinging over to him on the second ballot, adding six votes. He then had the required majority of forty-two, but asked for another ballot to confirm the vote and acquire a stronger sense of legitimacy. Whether in the third scrutiny he received forty-eight votes, the more likely result, or all sixty-one remains impossible to determine. Pius XI had strengthened the rigid secrecy of the conclave, when shortly after his election a cardinal's heirs sold his notes on the conclave of 1922 to a newspaper. He had ordered that henceforth all notes and records be burned with the final ballots. The conclave of 1939 had taken less than a full day, and when the white smoke rose above the Vatican, no one had any doubts that only Pacelli could have been elected so soon. What limited information participants provided about this conclave afterward indicates that it was entirely uneventful.

Chosen pope on his sixty-third birthday, Pius XII came from a Roman family that had provided papal officials for over a century. He was the first pope born in Rome since Clement X's election in 1670 and was educated for a career in the church hierarchy. After spending ten years working on the new code of canon law, he was sent in 1917 as nuncio to Bavaria. Sister Pascalina, a German nun, became his housekeeper in Munich and remained with him for the rest of his life. She became so indispensable to him that she was allowed to serve as his conclave servant in 1939, the only woman to have had that position. Late in his life she was the only person he trusted. In 1920 Pacelli became nuncio to the new German republic. He remained in Germany until 1930, while he became fluent in German and German culture. He was named a cardinal in 1929 and then secretary of state at Gasparri's death. His election made him the first secretary of state to gain the tiara since 1667.

Pacelli became Pius XII to honor his predecessor and his commitment to peace. Five months later the world was back at war. Shortly after it began, he signed a letter of resignation, which was to be made public if the Vatican were occupied. The cardinals who could were to assemble wherever they could freely elect a new pope. Pius's conduct and policy during World War II have become the focus of a major historical controversy, arising largely out of the effort to have him canonized. He found it very difficult to condemn the Axis powers when the Soviet Union, in his mind the greater danger to the Church, would have benefited from such an act. No one can deny that Pius could have been more forceful in denunciating Nazi crimes, especially the mass murder of Europe's Jews, about which he was well informed by 1943. On the other hand his critics ignore or reduce to insignificance such acts as his support for a plot against Hitler in 1940[10] and protection of Roman Jews, which led the chief rabbi of Rome to call him the best friend the Jews had. The title of a recent book, *Hitler's Pope*, is slanderous, but the case for sainthood would be far stronger had he risked more for Hitler's victims.[11]

Pius XII was the first pope who had traveled extensively outside of Europe before his election. It was reflected in his pattern of granting red hats. In his reign of nineteen years and eight months, he promoted fifty-six men to the College, a low number per year (2.9) compared to previous popes since 1878 (4.3). Pius made only two sets of promotions, waiting until the war ended to make his first, of thirty-two men, in early 1946, leaving the College one short of seventy. It was the largest single promotion to that time. Only four were curia officials; the rest were residential bishops. His second took place in 1953. Thirty-six of his creations (64 percent) were Europe-based; Italians constituted a quarter. All five other continents were represented among the others. Africa gained its first cardinal, albeit a native of Portugal. Three were Asians; nine were Latin Americans representing seven different countries; and six were from North America—two Canadians and four Americans. Pius said at his second promotion that he would like to create more cardinals but felt obliged to respect the tradition of seventy that dated from Sixtus V. Among those promoted was Giuseppe Siri of Genoa, whom many expected to succeed Pius, but missing from the list was Giovanni Montini, who for a decade had been the second-most-powerful man in the curia. Pius's refusal to give him a red hat indicated that he did not want Montini to succeed him. Pius's promotions to the College were largely responsible for the immense changes that have taken place in the Church since his death, which belies his reputation as a conservative. He once suggested that he was interested in having a non-Italian successor.[12]

When Pius died on October 9, 1958, he left fifteen vacancies in the College. He had intended to make another large promotion, but ill health in his last year prevented him. Two cardinals died right after him, and two from Communist states were not allowed to travel, so fifty-one attended the conclave, the smallest number since 1831. A third of the electors were Italian, the lowest percentage since the sixteenth century, but they were still the largest bloc. Like his predecessor, Pius never assembled all the cardinals together, so the non-Italians knew few of their fellows, and the Italians knew few of them. Twenty-four of the cardinals, almost half, were eighty years or older, since Pius had given red hats mostly to men who were over sixty. Since he had left the office of *camerlengo* vacant for years, Cardinal Tisserant, the dean, had to perform the duties associated with a pope's death, and the College elected the *camerlengo* to organize the conclave.

For the first time a pope's funeral and a conclave's opening ceremonies were televised around the world; so too were news accounts of what was rumored to be going on in the Vatican and views of the smoke rising from its chimney. Air travel allowed the non-Italian cardinals to reach Rome well before the conclave began, showing that Pius XI's rule requiring a fifteen-day delay was no longer needed. This allowed plenty of time for the *prattiche*. Every morning the cardinals met in a general congregation to plan the conclave and deal with church business that could not wait for the new pope, and after that there were long lunches, teas, and dinners in small groups to discuss the *papabili*. It was perfectly permissible in the *prattiche* to promote candidates or object to them, based on perhaps the vaguest rumors, but any effort to solicit votes for oneself was forbidden.

There was some support for Montini, who had been sent to Milan without the red hat associated with that see. Traditionalists thundered loudly against electing someone outside of the College. Siri, the curia's candidate, allegedly became so agitated by the talk of electing Montini that he slammed his fist on a table and smashed his episcopal ring. There was also talk of electing a non-Italian. Cardinal Agagianian from the Uniate Armenian Church, who had spent most of his career in Rome, had broad support. The bookmakers gave him odds of 3 to 1 along with Cardinal Ottaviani. Another non-Italian, Stefan Wyszynski of Warsaw, at 4 to 1, had the same odds as

Siri. The favorite of the betters, with odds of 2 to 1, was Angelo Roncalli, the patriarch of Venice.[13] Although he was almost seventy-seven years old, half of the College was older, and he was hearty for his age. He was aware that he was *papabile* but wrote to his sister before the conclave, "I am more than ever confident, almost certain, of returning to Venice."[14] Yet he prohibited his nephew from coming to Rome, which made sense only if he thought he might be elected.

When the cardinals entered the Vatican on October 25, 1958, they heard the traditional sermon on electing the pope, in which Monsignor Bracci, describing the sort of pope the cardinals should consider electing, came close to identifying Roncalli. Since Bracci was a member of the curia, it has been debated ever since whether he was speaking for himself or for the curialists in calling for a caretaker pope, as Roncalli was expected to be. The text of the sermon was published, but beyond that, the details of the scrutinies remain vague, as the ban on revealing the secrets of the conclave has been well observed. Roncalli would later tease reporters that while their efforts to learn the conclave's secrets were remarkable, the cardinals' silence was even more so.[15] His diary, published after his death, provides some insight into the election but reveals little detail. It is known that some conservative cardinals were eager to prevent Siri from being elected, not for his views but his age, since at fifty-two he might reign longer than even Pius IX. (Siri died in 1989; had he been elected in 1958, he would have been weeks short of Pius's record.) They turned to Roncalli as a transitional pope. The six French cardinals favored him, as he had been influential in getting red hats for several of them.

Only fragments of information have come out on the conclave's events. In the four ballots of the first day, Roncalli and Agagianian were the strongest candidates; in Roncalli's phrase they "went up and down like chickpeas in boiling water."[16] Montini consistently received two votes. In the fourth Roncalli had twenty votes to Agagianian's eighteen. Roncalli opposed Agagianian on the grounds that if the cardinals wanted to elect a non-Italian, they should not choose someone who had spent nearly his entire career in Rome. On the second day, October 27, Roncalli forged ahead to thirty-two votes, as the Armenian's support faded. That night it was clear enough that he would be elected, and cardinals came to his cell encouraging him to accept the office, making amends for past slights, or discerning what his policies would be. The next morning Ottaviani swung several votes to him, and in the third scrutiny that day, Roncalli received enough votes to become pope. The best authority on the conclave indicates that he had thirty-eight, three more than required. Siri received ten, and Montini, two. Roncalli later indicated that his vote had gone to Cardinal Valeri, his predecessor in Paris.

Roncalli, born in 1881 in the Veneto, came from probably the humblest background of any pope. His parents were peasants whose house shared living space with their animals, although they moved to a better house when he was nine. The priesthood was the way out for many a peasant boy, and he made that move by being ordained in 1904. He became secretary to the bishop of Bergamo and later taught in the local seminary. When Italy entered World War I, Roncalli was drafted and served in the medical service. In 1920 Benedict XV brought him to Rome, and after five years in the curia, he was appointed apostolic visitor to Bulgaria, an Orthodox country. Nine years later he was moved to Istanbul, where he dealt not only with the Orthodox Church but also four Uniate churches and the Muslim majority. His later interest in ecumenism came out of those experiences. Just when it seemed that he would always remain a minor papal diplomat, he was named nuncio to the French government being reconstructed after liberation in 1944. After a successful nine years in Paris, Roncalli was named patriarch of Venice and given a red hat. The five years he served there gained him a reputation as a fine pastor to go with his standing as a diplomat, and made him a *papabile* in 1958.

If Roncalli had any doubts, they had fully dissipated by the last scrutiny. He calmly answered that he accepted the office. When asked what name he would use, he pulled a sheet of paper out of his pocket with a speech on it. If his election was not a surprise to the College, his choice of name was: "I will be called John." It was surprising, because the name had not been used since the Great Schism. By becoming John XXIII he condemned the prior holder of the name as an antipope. When the Sistine Chapel's door was opened to allow an ill cardinal to leave, the new pope's secretary could see that the canopies over the chairs of all the other cardinals had been closed while his remained open, confirming that Roncalli had been elected. A rumor that the conclave was over had reached the huge crowd in St. Peter's square, but it took seventy-five minutes after the scrutiny had concluded for the white smoke to waft from the chimney. Several times during the conclave the smoke had been deceptively light at first, and the crowds had been the more disappointed by having been given false expectations. About an hour later Cardinal Canali appeared on the balcony to announce the thrilling news: *Habemus papam!*

The news of John's election was a surprise to the world at large, if not to Romans in the know, since he did not look or act much like the popes since 1914. He was fat and jovial, not lean and austere. He was not a rebel despite what he did as pope; cardinals do not elect rebels as pope. He retained most of the curia officers from his predecessor. It has been argued that keeping them was the price the curialist cardinals had required for their votes. John always insisted that he was following through on policies begun by Pius XII. Yet a new atmosphere was clear from his first public words as pope, when he spoke of unity with all Christians and peace in the world, without denouncing heretics, schismatics, and Communists, as Pius routinely did. A week after his election John decided to fill the College. He wrote in his diary that as he and Domenico Tardini, his secretary of state, began drawing up a list, which Tardini himself and Montini headed, they quickly reached seventy. Reflecting that when Sixtus V had set the limit, the Church was only a third as large, they added more. With twenty-three new cardinals, the College had seventy-four members. They were an international group, including the first native African cardinal and the first ones from Mexico, the Philippines, and Japan.

His secretary later insisted that John never made the famous remark about opening the windows of the Vatican and letting some air in, but that is what he did. He opened not only the windows but also the doors, as he left the palace to do pastoral work in his diocese of Rome, visiting hospitals, orphanages, and a prison as well as churches. The best-known act of opening windows was, of course, calling a general council. The idea of a council was not entirely novel. Pius XII had talked of convening one in the years after 1945, but his council would have been largely involved in defining such doctrines as the Assumption, which he did on his own in 1950. Several cardinals would later relate that they had spoken to John during the conclave about calling one, although his biographers downplay their influence on the decision. They insist that he came to the idea of the council on his own within days of his election. On January 25, 1959, John announced to a group of cardinals that he intended to convene a council soon, since his age would not allow for the six years of preparation that Vatican I had required.

This is not the place for an examination of either Vatican II or the politics and work that went into preparing for its opening on October 11, 1962. Its broad renovation of the Church did not touch the conclave, as there was no formal discussion of the papal election. A month before it began, John was diagnosed with stomach cancer, and at his age of eighty-one, there was no thought of surgery or radiation treatment. As was typical for popes, there was no public notice of his illness, and he was determined to see the council underway and, he hoped, finished without it being distracted or derailed by news that he was dying. The council's first session lasted until December 8. It was officially suspended to allow time for work on documents, and would resume in nine months; but curia officials who knew of John's illness did not want the council in session when he died, to ward off any possibility of the council playing a role in the coming election. In February 1963 he expressed worry about the conclave, fearing that the forces opposed to the council would elect a pope who would not continue his work. Shortly before his death, however, he was optimistic that Cardinal Montini would be chosen, although a dying pope's opinion on his successor has rarely had much impact.

John continued to expand the College, so at his death on June 3, 1963, it consisted of eighty-one members along with three in Communist lands whose names remained *in petto*. Twenty-nine were Italian, just over one-third, and another twenty-eight came from other European countries. The twelve Latin American cardinals constituted the next largest bloc, followed by the seven North Americans. For the first time a native African participated. Because of the speed of air travel, all the cardinals except for Montini, who remained in Milan until two days before the conclave, were in Rome within ten days of John's death. The rational for waiting further no longer held, but the delay remained in effect. Constructing rooms in the Vatican for eighty-one cardinals also required time. That gave the factions plenty of time to promote their candidates. Those described as "innovators" had two *papabili*—Montini and Giacomo Lercaro of Bologna. Lercaro had a better claim to John's legacy as the people's pope than did the shy Montini, but that only made those who wished to end or redirect the council more determined to prevent his election. He made the gaffe of writing to the Bolognese not to expect him back. That sort of thinking has never been rewarded with the tiara. The conservatives' favorite was Siri of Genoa, but aware that most non-Italians opposed him, which made his election impossible, they promoted Cardinals Confalonieri and Antoniutti, believing that their moderate positions on several conciliar issues would allow some innovators to turn to them when their own candidates proved unelectable.

The American CIA station in Rome reported Siri as the conservative most likely to win and Montini, most probable among the liberals.[17] One suspects that similar reports appeared in nearly every world capital. The CIA also reported that it was expected that New York's Cardinal Spellman, given a red hat by Pius XII, would vote for Siri, as he probably did in 1959, since they had known each other for four decades. Spellman had also worked with Montini in the curia in the 1930s and decided he was more likely to win. Montini's lingering in Milan caused some grousing among his supporters, for it meant he was not taking part in the *prattiche*. When he finally came to Rome, he and Spellman had lunch, and when it was over, he had Spellman's promise to vote for him. Spellman's pledge likely brought with it the vote of James McIntyre of Los Angeles, his protégé. The interest of the CIA in the papal election reflected the view that the papacy was a powerful ally in the fight against Communism. The cardinals were well aware of such interest from the CIA and the intelligence agencies of all the major states. They also knew that the media, including television, which was a major presence in St. Peter's square for this conclave, would do anything to get first word on the election's outcome. Consequently the conclave rooms were swept for listening devices and other means of secretive communication. There is no word that any were found, but there is a story of a conclavist who had a tiny radio he used to alert immediately either the CIA in one version, or Vatican Radio in another, of the result, flushing it down a toilet after using it.

On the afternoon of June 20, before the College was locked in, a conservative gave the pre-election sermon to the cardinals. He seems to have forgotten that John XXIII had appointed many of the cardinals present. His attack on John's legacy and ridicule of the late pope's simplemindedness alienated several cardinals who were less than pleased with the council's direction but owed their red hats to him. How the cardinals voted, however, is not known, since the veil of secrecy over the balloting has been more complete than for previous conclaves, and the best source for this conclave can only give estimates on the balloting for the candidates.[18] In the first scrutinies on the morning of June 21, Montini received about thirty votes, while both Lercaro and Antoniutti had around twenty. Cardinal Suenens of Belgium, perhaps the most liberal cardinal, asked that those who were voting for Lercaro move to Montini. In the day's third ballot Montini picked up about ten votes, while Lercaro lost the same number, but he was still some fifteen short.

Then, as preparations were being made for the fourth scrutiny, Gustavo Testa, to whom John had given a red hat in his last consistory, blew up at the two conservative cardinals who sat on either side of him; they had been whispering across him about strategy for the next vote. In a voice loud enough to be heard by all, he rebuked his neighbors for their dishonorable plotting against Montini and demanded that they think of the good of the Church. Siri in turn lost his temper and expressed outrage that Testa had violated the conclave rule against discussion. When calm returned, and the next ballot was counted, Montini had gained enough votes to be within several of becoming pope. There had been elections in the past when someone in Montini's position had failed to get the last needed votes, but it would not happen this time. The next morning Montini passed the required majority of fifty-five, but apparently only by two. Siri remained adamant against voting for him, as did some twenty other cardinals. Lercaro also held on to several votes to the end. Despite the efforts of the conclave marshals to ensure secrecy, the CIA received word of Montini's election within minutes.[19]

The election's outcome was no surprise to Montini, who announced that he would be Paul VI. It was the first time the name was used since 1621.

The new pope said that he had chosen it to show that he wanted to reach out to the Gentiles as Saint Paul had. Born in 1897, he was the son of the editor of a Catholic newspaper in Brescia who met his wife while both were on pilgrimage to Rome. The future pope had had a heart condition that prevented him from pursing a career in journalism. The priesthood being deemed less strenuous, he was ordained in 1920 and soon enrolled in the school for papal diplomats. He served in Germany and Poland and returned to Rome to work for Secretary of State Pacelli. When Pius XII's only secretary of state died in 1944, he refused to name another, and Montini and Tardini divided the office's responsibilities as undersecretaries. In 1945 an American intelligence document referred to Montini as Pius XIV, expecting that he would be the second pope after Pius XII. In 1954 he was made archbishop of Milan but without the red hat that always went with the see. Rumor said that he had fallen out of favor because of his dissatisfaction with Pius's hard line against Communism. Many observers felt that he would have become pope in 1958 had he been a cardinal. John named him in his first consistory.

Seeing Vatican II to its conclusion occupied most of Paul's time in the first two years of his reign, but he showed his commitment to being the new Paul by flying to Jerusalem, Bombay, and New York during its recesses. He thereby became the first reigning pope to travel outside of Europe. He invited non-Catholics and even non-Christians as observers to the council. Before the council closed on December 7, 1965, he and Greek Patriarch Athenagoras lifted the mutual excommunications of 1054. Paul expanded greatly the dialogue with Protestants and non-Christians that John began. He was the darling of liberal Catholics for several years as he put into effect the reforms of Vatican II, but his 1968 encyclical *Humanae Vitae,* reaffirming traditional doctrine on birth control, recast him in their eyes as a reactionary.

Like virtually every pope with a long reign, Paul became more conservative as his years mounted; but he continued to make major changes in the Church, including in the papal election. Paul was intrigued by Cardinal Suenens's proposal that, if not all the bishops, then the presidents of the national conferences of bishops (formed to give reality to Vatican II's expression of collegiality among bishops) should elect the pope. Siri took credit for persuading him not to make so drastic a break with tradition. In 1970, however, Paul took to heart Suenens's argument that the church hierarchy was a gerontocracy, by requiring bishops to retire at age seventy-five and removing the right to vote from cardinals over eighty, who also had to retire from service in the curia. It is also possible Paul realized that nearly all of the cardinals over eighty had not voted for him as pope. The document did not bar a cardinal over eighty from being elected. Ten cardinals lost their vote, and they were furious, especially since most belonged to the conservative wing of the College and it appeared that Paul was trying to rig the next conclave.

Paul reiterated the strict rules for secrecy at the conclave, replaced the two conclavists for each cardinal with a set of secretaries and servants, and included for the first time prohibitions against bugging devices.[20] He also made the position of dean of the College elective from the cardinal-bishops instead of being based on seniority, thereby reducing the possibility that the dean would be over eighty and ineligible to preside over the conclave. The joyful task of announcing the name of the new pope to the world was given to the senior cardinal-deacon. When the pope turned eighty in 1977 without abdicating, the grousing among those who had been disenfranchised grew louder. Besides the bad results of Celestine V's resignation, the principal argument against papal retirement is that a retired pope, despite his best intentions, would create a faction against the new pope by his very existence; the situation would become more serious if the new pope changed policy, as often happened.

Paul went well beyond John in expanding the College. He promoted 120 men; a small majority came from outside of Europe, although many were ethnic Europeans. At his death on August 6, 1978, there were 130 cardinals, but 15 were over 80 and ineligible to vote because of Paul's edict, leaving 115 voters. One died before the conclave began, and three were too ill to travel. Of the 111 who attended, 100 were Paul's creations. Only three remained from Pius XII's reign, and eight came from John XXIII's. Fifty-five cardinals were seventy years of age or older. Paul's deteriorating health was a better-kept secret than John's had been in his last days, so his death came as a surprise to most cardinals. Although air service brought the non-European cardinals from all corners of the world faster than in 1963, preparing cells in the Vatican for so many electors took longer. Finding workmen to do the labor proved difficult because it was during the Italian vacation time. It took nearly twenty days to finish the work. The most important reason for the delay, however, was that the cardinals over eighty persuaded the others that, since they could not vote themselves, they deserved time to make their views known to the electors in the *prattiche*. Several sought unsuccessfully to have Paul's decision overturned.

Most cardinals had attended Vatican II as bishops, and the more open system of church governance it created also meant they knew one another. Cardinal Confalonieri, who had served as a conclavist in 1922, was the dean, but he was too old to participate under the new rule. Jean Villot of France, the secretary of state, was the *camerlengo,* the first non-Italian to hold the office. He had little chance of being elected, since he was blamed for the long delay in beginning the conclave. He had used the telephone to inform the cardinals of Paul's death, and Siri and others insisted it be done properly, by letter. Although the College was three-quarters non-Italian, there was less expectation than in 1963 of the election of a non-Italian pope. While the cardinals discussed the candidates, for the first time there were public campaigns to influence them. Followers of France's Archbishop Lefebvre, who refused to accept Vatican II, announced at a press conference that his group would not recognize an election that excluded the old cardinals. Civilita Cristiana, a conservative Italian group, distributed flyers that called for a "Catholic pope." The Committee for the Responsible Election of the Pope, mostly liberal Americans, demanded public balloting. Some theologians published an open letter in a Catholic newspaper with a job description for the pontiff, which emphasized ecumenism.

Bookmakers in London gave odds of 5 to 2 for Sergio Pignedoli, 7 to 2 on Sebastiano Baggio and Ugo Poletti, and 4 to 1 on Carlo Benelli. The best odds on a non-Italian were on Dutchman Johannes Willebrands, at 8 to 1. With so many electors, the factions were many, although their memberships were not clearly defined. One was the party of conservative Italians largely associated with the curia; the curialists numbered only twenty-eight, since Paul had given red hats mostly to residential bishops. Their leader was Siri, and their strongest candidate was Baggio. Benelli, who at age fifty-seven was deemed too young, led the moderate Italians, who promoted Pignedoli with Albino Luciani of Venice as their second choice. The other Europeans, led by Franz Koening of Vienna, supported Poletti or, if the conclave could be persuaded to turn to a non-Italian, Willebrands. The Third World cardinals, whose leader was Brazilian Aloisio Lorscheider, preferred Baggio, since he had visited most of them as president of the Congregation of Bishops, or Lorscheider himself. A group of neutrals included the seven cardinals from the United States.

On August 25, after the Mass and the homily for the election of the pope, which was uncontroversial, the Vatican's doors were locked behind the cardinals after an exhaustive search for interlopers and bugging devices. A support staff of about seventy to replace the conclavists was also locked in. Paul eliminated the conclavists both to enhance secrecy and to accommodate the far greater number of electors in the conclave. Not all the cells were equally small, but they were all furnished alike, with a hospital bed and a thin mattress, a wooden chair, a kneeler, and a few toiletries. With all the doors and windows covered and little artificial light provided, the ancient rooms were dark, hot, and stuffy. Siri commented that being in conclave was like being in a tomb. But that was the method behind the madness, to persuade the cardinals to make a quick decision.

This first evening of the conclave the cardinals listened to another reading of Paul's constitution on electing the pope and swore again the oath of secrecy during and after the election. Paul's constitution banned discussions among the electors after entering conclave and even more sharply barred any deal making. One can assume that it was quieter in the Vatican that night than in previous conclaves. The next morning after Mass, the cardinals assembled in the Sistine Chapel for the first scrutiny. The traditional canopies over the heads of the electors had disappeared; the cardinals sat at small tables crammed together, which made it difficult to write their ballots without their neighbors seeing. The enhanced secrecy has made it difficult to discern the exact voting, but enough information has leaked out to provide some sense of what happened. In the first scrutiny, Siri had about twenty-five votes, Luciani, twenty-three, and Pignedoli, eighteen. Luciani's strength was unexpected but not unwelcome to cardinals from a wide range of factions.

The second ballot saw Siri's votes drop sharply, while Luciani shot up to near fifty, suggesting that the curialists regarded him as acceptable. Most accounts of the conclave agree that there were four scrutinies, but Luciani had already passed the needed majority of seventy-five by a vote or two in the third. He then asked for a fourth ballot in order to surpass eighty-five votes, to avoid any challenge that his election was invalid, because he had not received a vote of two-thirds plus one of all 130 cardinals including those over eighty. He received ninety-six. It is unclear whether a particular faction did not vote for him. After the conclave many cardinals emoted effusively about how obvious the working of the Spirit had been.[21]

Born in 1912, Luciani came from the mountains north of Venice. His father, a laborer, was a member of a socialist party for a time, while his mother was a seamstress. Luciani is the only pope to come out of the working class. John made him a bishop in 1958 and Paul gave him the see of Venice and a red hat in 1969. When asked what name he would be called, he replied that to honor the two popes, he would be John Paul I (strictly speaking, there can not be I until there is II). The new pope with a constant smile and a great sense of humor won the hearts of the world from the first moment of his papacy. In the case of John Paul, the term "reign" is inappropriate, because he jettisoned most of the trappings of the papal monarchy. "Coronation" being a term that he deemed obsolete, he refused to wear the tiara for his consecration. He abandoned the papal sedan chair but had to go back to using it when the people complained that they could not see him. He hinted at the papacy's budding financial crisis by cutting in half the gift to the curia staff at his election.

What few policy decisions he made in the next month indicated a true middle-of-the-road approach. Before he could begin to implement his program, however, he was found dead in bed early September 29, 1978. The official cause of death was a heart attack, and it came out after his death that he had had heart trouble for some time. Curia officials declared that an autopsy could not be done for a dead pope, which was based only on recent practice, as several popes had been autopsied in the past. The lack of an autopsy fed rumors that he had been poisoned by those involved in the burgeoning financial scandal centered on the papal bank.[22] Although it became known that curia officials misinformed the public on what the pope had been reading when he died (a stack of reports, not *The Imitation of Christ*) and who found him (a nun-housekeeper, not his male secretary), there is no evidence that they lied about the cause of death. After the fifth-shortest papacy in history, the stunned cardinals were called back to Rome for another conclave, the last of the twentieth century. It would elect a pope who, as of May 2003, has had the third-longest reign in history.

Chapter XI

And On to the Twenty-first Century

The anticipation and enthusiasm that had marked the conclave of August 1978 were little in evidence as the cardinals rushed back to Rome following John Paul I's sudden death. The cardinals were somber as they began meeting in the general congregations. The return to health of an absent cardinal offset the dead pope, so the cardinals who participated were again 111. The issue of the health of the *papabili* was far more at the forefront than in prior elections; the cardinals did not want to choose again someone who would die soon after his election. The quick end of John Paul I's papacy meant that the College would be inclined to seek the same sort of pope—a pastoral bishop who was not part of the curia—but a more vigorous one.

For the third consecutive conclave, Cardinal Siri was the most mentioned name among the conservative cardinals. Although outspoken in his rejection of much of Vatican II, he had been an activist pastoral archbishop in Genoa and had friends among the centrists. What chance he had at election disappeared upon publication of a newspaper interview in which he ridiculed the council and the idea of collegiality between the pope and the bishops. The newspaper had promised to hold it until the conclave doors had been locked, but when another interview by Siri appeared in a rival paper, the editor felt free to publish it a day early. Thus, Siri's views became known before the cardinals went into conclave, and many who were considering voting for him were shocked at his firmly reactionary views.

The College's liberal wing had no strong candidate. The leading liberals of earlier were still in place, but there was little discussion of them. It is not clear whether there was little more to say about them, or if it was felt that they would not be elected, having been passed over a month before. The most talk was about electing a non-Italian. Brazil's Aloisio Lorscheider was regarded as the probable choice should that happen, but those in the know were aware that he had undergone heart surgery only weeks earlier. He was well enough to attend the conclave, but there was no chance that he would be elected. Franz Koening of Austria, the most influential of the non-Italian Europeans, spoke often of Poland's Karol Wojtyla, but the possibility of a pope from a Communist state seemed too remote for reporters and pundits to pick up on the favorable comments on Wojtyla.

In the late afternoon of October 14, the cardinals processed into the Vatican. There was nothing unexpected about the arrangements for the conclave. After the August conclave there had been some leaks about the voting; therefore, Cardinal Villot, the *camerlengo,* stressed even more the obligation of secrecy. Thus, the vote counts for this election are among the least certain for any conclave. It is not known for sure how many ballots were taken, although the number of times black smoke issued from the Vatican chimney and the timing of the white smoke in the early evening of October 16 suggest eight. Nonetheless, scenarios on how the voting went have been produced. They contradict each other, and so far, there has been no confirmation of the accuracy of any of them. They are in general agreement that Siri received the largest number of votes in the first scrutiny, perhaps as few as twenty-three or as many as thirty-six. Carlo Benelli of Florence, previously a popemaker and now a candidate in his own right, came on strong in the next scrutinies. His fifty-seven years, a negative factor in August, were more positively viewed now. He gained votes rapidly, possibly coming within five of required seventy-five in the fifth scrutiny. Although he had long been a curia official before being promoted to Florence, most curialists opposed him, as did some liberals, who regarded him as too authoritarian. Adamant that Benelli not be elected, they spent the afternoon between the fifth and sixth scrutinies working to prevent his election. In the sixth scrutiny some votes moved away from him to other Italian *papabili,* but seventy-six-year-old Giovanni Colombo, the strongest among them, made it clear that he would not accept the office, because he felt an obligation to retire at age eighty. Later comments from two participants suggest that dramatic events occurred in the conclave. Cardinal Carberry of St. Louis said, "I would like to tell you everything. It would thrill you. But I can't." And Siri, the leader in the first several ballots, indicated that it would be a good thing to reveal what happened in the conclave at some point, "for secrecy, though valid at the time of the conclave, can hide some very uncharitable actions."[1]

Meanwhile, Wojtyla was gaining strength. He was well known to both the Italians and the non-Italians, having gone to Rome frequently since becoming a cardinal in 1967 and traveling extensively outside of Europe. Like Benelli, he was young for a *papabile* at fifty-eight, but his age and robust appearance now were advantages. He was a man of broad intellectual interests yet well-grounded in Catholic theology, a supporter of Vatican II, an activist pastoral bishop, and a skilled veteran of the political wars between the Church and Poland's Communist government. He also fluently spoke several languages, including Italian. The major hurdle was making the leap to electing a non-Italian. Once a cardinal jumped that hurdle, and for some it came as early as the first scrutiny (in which Wojtyla had five votes), he was the most obvious choice. Koenig used his influence among the European cardinals in his cause, and John Krol of Philadelphia, a Polish-American, brought the American cardinals over. Krol is said to have argued that the scandal in the Vatican bank required a non-Italian pope. By the sixth scrutiny, Wojtyla had about twenty-five votes. Before the next ballot, Cardinal Willebrands, who was receiving about twenty votes, told his supporters to stop voting for him. So did Cardinal Baggio, who had been strong in August but never passed twenty votes in October. Most in both groups went over to Wojtyla.

Exactly what happened in the next two ballots is the most uncertain aspect of this conclave. Two scenarios have appeared to explain the final results. One has it that Wojtyla received enough votes in the seventh scrutiny to become pope but asked for another ballot to confirm the decision. It proposes that while he was over the seventy-five needed, it was not by much. In the eighth scrutiny he was elected with perhaps as many as ninety-nine votes, but a dozen cardinals, allegedly Siri's supporters, refused to make the vote unanimous. The other scenario more simply has it that Wojtyla was close to victory in the seventh scrutiny and went well above seventy-five in the eighth, reaching perhaps the ninety-nine of the first scenario. Wojtyla had done nothing to win votes and had expressed anguish at the prospect of becoming pope in the hours after the morning ballots. Wyszynski, his senior in the Polish Church, emphasized to him both his obligation to accept the Spirit's will and the great honor his election would bring to Poland. When it was certain that he was elected, the dean asked him if he accepted, and, slumped over in prayer, he paused for a long time before saying, "I accept." Asked what name he chose, he replied that he would be John Paul II to honor his short-lived predecessor and Paul VI, calling Paul his inspiration and strength. One story says that before the eighth scrutiny he told his friends that he wanted to be Stanislaus, but Koenig sharply told him that he would be John Paul II.

As it was mid-autumn, it had been dark the night before when the ballots for the afternoon scrutinies were burned, and there was confusion whether the smoke was white or black. A phone call to the conclave marshal clarified that it was meant to be black, and the public address speakers gave the news to disappointed people in St. Peter's square and around the world. At the burning of the morning ballots on October 16, the smoke at first was white and then turned black, and again an announcement had to clarify it. At 6:18 that evening, however, as smoke began pouring out, there was no question that it was white. Within a half-hour, probably the shortest time ever, Cardinal Felici appeared on the balcony overlooking the square to announce *Habemus papam*. When he dramatically paused after saying Carolum, Latin for Charles, the well informed in the crowd knew that there were a dozen cardinals who had that first name, including Benelli. Felici then proclaimed Wojtyla. Few in the square recognized the name, but they knew it was not Italian. The wild cheering greeting every word of the announcement fell silent, as the crowd seemed stunned by the news. Over 450 years of Italian monopoly on the papacy had been broken. Many Italians in the crowd were not ready to accept that fact. Then the new pope appeared himself. Blonde, broad-shouldered, and robust, he looked like no pope that any of them had seen. (Pius X was the only pope in the past century who was close to matching John Paul II's physical appearance.) When he began to speak in fluent Italian, he soon had the people in his grasp. He won them over completely when he said, "Even if I cannot explain myself well in your—our Italian language, if I make a mistake, you will correct me." And he has been a far more active bishop of Rome than most Italian popes.

In the twenty-four years that John Paul II has been pope, he has demonstrated a deep respect for the traditions and authority of the papacy. As is natural for someone who has held a position so long, it has become more obvious in recent years. In several respects, however, he has broken with tradition. The combination of the length of his reign and the greater number of cardinals has allowed him to surpass the previous largest number of appointments in a reign by half. John Paul accepted Paul's rule that cardinals who turn eighty on the day before a pope's death or are older not serve as papal electors.[2] When, in February 2001, the pope named forty-four cardinals, the record for one consistory, it was pointed out to him that he had far surpassed Paul's limit of 120 electors; his only comment was a laconic, "Yes, I know." Two years later, however, 56 of the 167 cardinals have turned 80 and cannot participate in the conclave. With five more turning eighty in 2003, the next conclave could be smaller than those of 1978. The 111 now eligible to be electors come from 56 countries. Paul VI appointed five of them. Assuming that John Paul does not name any more before he dies, the Italians make up only 16 percent of the voters, while non-Europeans now constitute 53 percent. That will increase the odds of getting a pope from Latin America, Asia, or Africa in the next conclave. The odds of getting a pope older than John Paul was at his election are also high. As of May 2003, only five cardinals are under sixty; the youngest is fifty-seven, which probably makes for the most elderly College ever. Only three cardinals are younger than John Paul was at election, and all will be older by the end of 2003.

The next conclave will be different in key ways from previous ones. In early 1996, John Paul issued an apostolic constitution on papal elections, which, it is true, largely reaffirmed Paul VI's decree of 1975.[3] Ignoring continuing agitation that the presidents of the conferences of bishops also serve as electors, he began by identifying cardinals as the only voters. It also noted resignation along with death as ways in which St. Peter's throne can become vacant. This sparked speculation that John Paul would resign when he reached the age of eighty in 2000, but thus far it has not happened. The pope agreed with theologians and canon lawyers that there is no requirement beyond tradition for using the conclave as the format for electing a pope, but he mandated that it continue to be used with some modifications.

The new constitution requires that at the first general congregation after the pope's death or resignation, the cardinals swear an oath to observe the electoral constitution exactly and observe total and permanent secrecy about the conclave's events. Although nearly every point on earth is now accessible to Rome in a day's travel by air, the conclave cannot begin before the fifteenth day after the pope's death, and the College can delay it up to the twentieth day. It must begin by the twentieth day. Anyone the pope has named a cardinal in consistory has a right to participate, even if he has not received his red hat. No duly appointed cardinal can be denied the right to vote or be elected except those who have been properly deposed or have resigned, whom the College cannot rehabilitate during the papal vacancy. Nor is it permitted to make any changes in the format for electing the pope. If there is a dispute over interpretation of what the rules require, a majority of the College will decide the issue.

After mandating that elections must take place in Vatican City, John Paul made a major change for future conclaves. The cardinals will be housed in Domus Sanctae Marthae, a new structure about 350 yards from the Sistine Chapel on the far side of St. Peter's. Ordinarily a hostel for prelates with business at the Vatican, the cardinals will use it for the conclave. That will save the expense of preparing cells in the Vatican and substantially reduce a conclave's cost. The rooms are spartan but far more comfortable than the makeshift cells. The pope must believe that the original justification for the discomfort of the traditional conclave—to compel the cardinals to make a decision quickly—no longer holds. The last conclave that went over four days occurred in 1831, which lasted fifty days. The scrutinies will take place in the Sistine Chapel as before. John Paul laid down rigid rules on secrecy in the Domus and during the movement of the electors to and fro. They will either walk through St. Peter's or be taken by bus. The rules set out what personnel—confessors, secretaries, physicians, nurses, housekeepers, cooks, and waiters—are permitted in the conclave. They all are required to take the oath of secrecy under pain of excommunication.

Radical changes that John Paul made in the electoral procedures reduce the likelihood of a long, deadlocked conclave. The events of the day the conclave begins differ little from the past. In the morning the cardinals attend the Mass for Electing the Pope in St. Peter's and hear a homily on their duty to choose the right pope for the times. In the afternoon they then assemble in the Pauline Chapel in the Vatican palace and process to the Sistine Chapel singing *Veni Creator.* They swear an oath to follow rules set forth in John Paul's Apostolic Constitution and observe permanent secrecy unless the new pope specifically permits them to reveal information. The *camerlengo,* his three assistants, and two trusted technicians are given responsibility to ensure that there are no devices that can transmit pictures, voices, or any information from the Chapel. John Paul reflected his era by his emphasis on preventing electronic eavesdropping. After the oath, the master of papal liturgical celebrations shouts the famed *Extra Omnes,* and all non-electors leave except the master and a priest chosen to give a second homily on electing the pope. After the homily they depart, and the cardinals once again swear the oath. The dean or his substitute then asks if there is any reason not to proceed to the first scrutiny. If there are issues to be settled, they are decided by majority vote, and the cardinals then proceed to the first ballot, although it can be delayed to the next morning.

John Paul abolished election by acclamation and delegation. The former was the practice of the cardinals acclaiming a candidate as pope without a written ballot; the latter was the practice of choosing a small committee that would then settle on a pope whom all other cardinals would accept. Neither method has been used for over two centuries. This leaves only the written secret ballot as the means of electing a pope. The constitution emphasizes the secrecy inherent in a written ballot as a factor in mandating its sole use. It makes no mention of election by accession or of the obligation of a cardinal not to vote for himself. Two-thirds of the ballots, or two-thirds plus one if the total is not evenly divisible by three, is required for election. The insistence on two-thirds plus one so that a vote for one's self does not create the winning margin has disappeared. The voters do not put their names on the ballot. The ballot now only has the words "I choose for Supreme Pontiff" at the top of the ballot and the name of the choice at the bottom. The ballot is to be folded so that the name of the choice is not visible until the three scrutators open it. Should two ballots be folded together in such a way that indicates the same person folded them, they count as one vote if the name on them is the same; if different, neither vote is counted. Three cardinals, the *infirmarii,* are selected by lot to take a lock box for the votes of those cardinals so ill they have remained in the Domus Sanctae Marthae. The *infirmarii* will vote immediately after the senior cardinal to prevent a long delay. It is not clear how they will travel to the Domus and back, but their obligation of secrecy is strongly emphasized. All ballots and any notes cardinals have made are to be burned after the final count has been ascertained. There is no mention of using the smoke as a signal to those outside, which has in fact been true for every decree on the conclave. The *camerlengo* is obliged to write reports on the scrutinies, which will be placed in the Vatican archives and kept secret unless a pope releases them.

The College may proceed to a scrutiny the afternoon they enter conclave, but only one may occur that day. From then there are to be two in the morning and two in the afternoon. If after seven scrutinies no pope has been elected, balloting will be suspended for up to a full day while the cardinals reflect on the situation and the senior cardinal-deacon exhorts them, an element already present in Paul VI's constitution. Then balloting is to continue for seven more scrutinies. Should there be no outcome by then, another pause will occur, with the senior cardinal-priest giving the exhortation. Following another fruitless seven scrutinies, a third pause will occur with the senior cardinal-bishop exhorting his colleagues. If after seven more ballots, there still is no pope, the College must deliberate on what to do. By majority decision it can choose to continue in the same manner, or reduce the required majority to one more than half of the voters, or vote only on the two men who in the preceding scrutiny had received the greatest number of votes. In this last format a majority of one is needed to elect a pope. If enough cardinals are adamantly opposed to both final candidates, they could continue to stalemate the election by refusing to vote for either. John Paul's constitution does not forbid cardinals from writing "no one" on their ballots.

The new procedure could dramatically change the dynamics of papal elections. If a group of cardinals constituting close to a majority of the electors is adamant about getting its candidate elected, it can simply hold out for ten to thirteen days until the number needed for victory may change. Had this procedure been in effect in the past, the history of the papacy would have been far different. There probably would have been a Pope Pole, for example, and not a Pope Carafa. Clearly the political situation for the papacy in the early twenty-first century is far different than in the mid-sixteenth, when monarchs interfered openly in the elections, but there is no certainty that similar circumstances will not reappear in the future. One can be optimistic that cardinals will always be motivated to vote only for the person best suited to govern the Church, but that job description can be interpreted very differently by different individuals, and the College might be bitterly divided. A protracted election that brings into play the new rule on reducing the required majority would reveal that the College was indeed badly divided. Will a pope elected by a simple majority of the cardinals instead of the traditional two-thirds hold the loyalty of the large number of Catholics likely to be disappointed in such a choice? If the two-thirds rule is waived, a significant part of the Church might well refuse to accept a pope elected in that way. Archbishop Lefebvre, his followers, and a few other Catholics rejected the popes elected in 1978: The absence of the cardinals over eighty made the elections invalid. That consideration might well persuade the College not to exercise the option of using the simple majority, especially since the conditions of the next conclave will not be as severe as in previous ones.

Much of the fascination of the next several conclaves will involve seeing how the new procedures play out, assuming the next pope does not use his prerogative to change them again. The changes wrought by John Paul in the conclave are almost revolutionary in an institution so controlled by tradition as the Church has been. Only time will tell whether they will benefit it.

The system that has been in place has well served the Catholic Church since the end of the Great Schism. There has not been a serious challenge to the legitimacy of any pope since Martin V was elected at the Council of Constance. No other system of governance comes close to matching that record. After every conclave, some Catholics have been disappointed in the choice of pope, but the number of those who have refused to recognize his legitimacy has been minuscule.[4] Too much tinkering with the electoral system would inevitably call it into question, as happened on a small scale in 1978 when the elimination of the post-eighty cardinals led some to challenge the standing of a pope elected with less than two-thirds majority of the full College. The rigid secrecy that has been enforced since 1903 can be perceived as an innovation, at least in contrast to the lax enforcement of the rules on secrecy in previous centuries, but it still lies within the parameters of conclave rules since 1274. It can be argued that the quality of popes since 1903 has improved, but that is probably more a result of the absence of the blatant political interference found in prior eras, than of the requirement of absolute secrecy.

With the current pope past the age of eighty and feeble in body albeit not in mind, speculation about who will be that next pope has been rampant. One consequence of Paul VI's exclusion of cardinals of eighty years or more is that it makes prognosticating the results of the next election before the pope dies even more difficult. Not only will deaths of cardinals change the makeup of the College, some cardinals also will turn eighty by then. A cardinal's obligation to retire at eighty will make it less likely, it would seem, that a man close to that age would be elected, because of considerable pressure on him to resign when he reached that age. Cardinal Laghi, whose influence in the curia would have made him a strong candidate for the conservative cardinals who are the majority at present, has turned eighty. His chance of becoming pope appears to have been greatly reduced.

John Paul II has greatly changed the ethnic makeup of the College in favor of non-Italians and even non-Europeans, but he has also strengthened the clout of the curia by naming many non-Italians to the presidencies of the congregations. Before his February 2001 promotions, it seemed that he was preparing the way for a non-Italian curialist to be elected, but his last set of red hats was skewed in favor of residential bishops. Their views also are less conservative than those of his previous promotions. The case has been made that John Paul's reign will result in a reaction against a non-Italian pope and the election of an Italian, in the same way that Adrian VI's policies ensured that an Italian would follow him.[5] That scenario, plausible in some ways, fails to take into account that Adrian's brief reign prevented him from changing the College. He named only one cardinal, who did not attend the next conclave, leaving the choice of a successor entirely in the hands of those who resented his attempts at reform. As of May 2003, John Paul has named all but five of the electors. While cardinals rarely have been clones of the popes who appoint them, there is no reason to believe that most cardinals are so upset at John Paul's policies that they would refuse to consider a non-Italian again. The odds are high, however, that the next pope will not be a Pole, simply because the current Polish cardinals lack the stature to be *papabili*.

The October conclave of 1978 broke a tradition dating from 1522 by electing a non-Italian. Will the next one end the other long-standing tradition begun in the same year by selecting someone not present at the conclave? It is most improbable that any cardinal will be absent except for serious ill health or imprisonment, and no one in either situation will be elected. Going back another 144 years to 1378, could a noncardinal become pope? The leap to a noncardinal is much greater than the leap to a non-Italian cardinal. The odds are huge that a cardinal will become the next pope, since the increase in the College's size has reduced the number of noteworthy prelates without red hats who might garner the attention of the voters. The odds against a cardinal from the United States being elected have always been high, but the recent sex abuse scandal within the American church appears to have made it impossible for an American to become the next pope. That is especially true of Bernard Law, previously regarded as the American with the best chance, who was forced to resign his see of Boston, although he is still a cardinal.

Predicting who will be the next pope can truly begin only when the pope has died, as cardinals die or reach the age of eighty before then. Each one who does not participate in the conclave changes the calculus of the next election. For example, when Laghi turned eighty in 2002, not only were his chances of being elected much reduced, but his influence and skill as an ecclesiastical politician will be removed from the next conclave. If still alive, he will make his influence felt in the *prattiche,* but should the conclave dead-lock, his faction will surely feel his absence in the ensuing negotiations.

Who will be the next pope? The prophecies of St. Malachy (the twelfth-century Irish monk, attributed with a set of pithy sayings that have allegedly predicted the last 112 popes, but which were probably created in the six-teenth century) proclaim that the next pope will be *de gloria olivae,* " on the glory of the olive." Does that mean that Cardinal Carlo Martini, archbishop emeritus of Milan, will be elected? He is a noted theologian, centrist in his views, and an admired pastoral prelate. Certainly he will be regarded as *papabile* should the next conclave occur soon, since he turns eighty in 2007. Whether one sees St. Malachy's prophecies as authentic or fraudulent, he has had as good a record in predicting popes as any pundit, journalist, bookie, or historian.

APPENDIX

Sfumata!

"Fumo! Fumo!" The huge crowd in St. Peter's Square shouts in one voice. The people know from the small cloud of smoke that wisps out of the tiny ancient chimney on the roof of the Vatican Palace that the cardinals sequestered within have completed a round of balloting in the process of electing the next pope. If it is black, the balloting has not resulted in the selection of the pope; if white, then it has. The *sfumata* is by far the best-known part of conclaves. Ask anyone in the Western world to name some element of a papal election, and the odds are very high that the use of smoke as a signal denoting the status of the election will be the one mentioned. Yet the history of its use is also perhaps the most uncertain aspect of the history of the conclave. None of the papal constitutions on the conclave mentions the use of smoke for signaling the crowd outside, although those from after 1500 require burning the ballots. Ludwig von Pastor's monumental work on the papacy takes no notice of smoke for the forty-three conclaves from 1417 to 1775 that he covers, often in great detail. Speaking of the election of Innocent X in 1644, he wrote, "The thunder of the guns of Castel S. Angelo and the clanging of the bells of the city proclaimed to the Romans that St. Peter's Chair was again occupied."[1] Pastor wrote his volumes during the pontificates of Benedict XV and Pius XI, when the use of smoke as a signal was well established, so presumably he would have been alert to the practice and have noted it if it was mentioned in his sources.

The first mention of ballot burning came from the conclave at the Council of Constance in 1417, which elected Martin V, although the context of the work suggests it was an established practice by then.[2] Burning probably appeared close to the time that the written ballot was adopted, which was definitely in use by 1294. Many elections in the early modern era were done by adoration and acclamation, however, which would have not produced any written ballots. When the conclave moved to the Vatican Palace, the ballots were burned in a small stove, but because the smoke was damaging the frescos in the Sistine Chapel, a special pyramid-shaped stove was introduced in the early 1600s to reduce the smoke. There was no chimney to the outside, which was added only in the eighteenth century.[3] With Gregory XIV's constitution of 1621, which banned adoration and acclamation and required two scrutinies a day, the amount of smoke from the burning ballots along with the candles and torches would have been significantly more noticeable, especially during the long conclaves after 1650.

According to a description of the election process written after the conclave of 1667, "the last act of the Post-scrutiny consists in the burning of all the ballots, which belongs to the function of the *Scrutators,* who cast them into the fire in the presence of the whole College."[4] There is no notice of using wet straw to produce black smoke or signaling those outside with the smoke. The same source related that the first indication that the conclave had ended was "the breaking of the walls that shut up the Loggia of the Benediction over the portal of the Church. There the principal Cardinal-deacon goes and cries out to the people 'Vivat N. who is made pope and has assumed the name of N.'"[5] A conclavist's diary from the conclave of 1775 noted that the cardinals burned the ballots, but made no mention of the smoke informing the crowd of the conclave's progress. It appears that firing the cannon was the first sign that a pope had been elected.[6]

The first notice of the *sfumata* acting as a signal that I have found comes from the election of 1823. A nineteenth-century study of Leo XIII's election stated:

It is a mistake to believe that once the pope is chosen, there is no smoke. Smoke is produced even when the pope is chosen, in order to preserve the secrecy of the votes. . . . Of the three instances of smoke in the conclave of Leo XII [1823], the first was seen distinctly, the third barely. The second was not, because it occurred at night. The smoke emitted in the third instance erroneously persuaded the people that the pope had not yet been chosen, but the experienced were not misled and waited a bit longer. Those involved in the conclave and in the ceremonial have a more secure indication that the election has occurred. If, when the smoke has appeared, the cardinals leave the chapel immediately, the pope has gone up in smoke. If they remain, the pope has been chosen.[7]

While that author indicated that the experienced knew that the sighting of smoke did not necessarily signal an unsuccessful ballot, other sources on the nineteenth-century conclaves took for granted that there would be no smoke visible after the final scrutiny.

A biography of Pius IX published in 1877, before the author was conditioned to think in terms of white and black smoke, related:

At each unsuccessful ballot the papers were burned, and the blue smoke escaped from a flue at a well-known spot [above the Quirinal Palace]. It appeared at a regular hour morning and evening, each time informing the expectant crowd that there was no election. On the 16th every eye in the multitude watched for the bluish column of smoke as the critical hour drew nigh. The hour passed, and the hateful signal appeared not. People began to breathe more freely. . . . At length becoming satisfied that some decision had been arrived at, they set up a mighty shout. . . . Soon were heard the blows of hammers breaking open a [boarded] window.[8]

Accounts of the conclave in 1878 also indicated that the *sfumata* had the function of signaling an unsuccessful ballot. They noted the yellow smoke that curled out of the chimney atop the Sistine Chapel and the crowd's reaction to it. The New York *Times's* story on Leo XIII's election stated, "The smoke of burning ballots having been seen at 12:30 PM today, the crowd before the Vatican, thinking that the ballot was again without result, had almost dispersed, when at 1:15 PM Cardinal Caternini appeared on the Grand

Gallery."[9] A description of conclave procedure written prior to the 1903 election described the 1878 conclave's final moments, "Cardinal Borromeo, whose duty it was to burn the voting papers, had been careful to use no straw after the last scrutiny. Nevertheless, the *sfumata* was sufficient to be seen in the Piazza, where the assembled crowd took it as a sign that the election was not yet made and returned to their homes for a few hours until the time for the afternoon scrutiny should arrive. Consequently there were but few people to hear the announcement made by the aged dean." Another source stated that "the smoke is a sign that no election has taken place."[10] Such statements attest that visible smoke signaled an unsuccessful ballot, and that the burning of the final ballots was not expected to give off enough smoke to serve as a signal.

At the 1903 conclave the meaning of the *sfumata* apparently remained the same. The newspapers indicated that when the *sfumata* failed to appear at the expected time in the late morning of August 4, the crowd recognized that the election probably had been made. There was no smoke seen for the final ballot.[11] The 1908 edition of the *Catholic Encyclopedia* has this on the smoke, "When there is no election, straw is mixed with the ballots to show by its thick smoke (*sfumata*) to those waiting outside that there has been no election." There is no mention of white smoke signifying the conclave's conclusion.

The first election in which the appearance of white smoke clearly signaled the conclave's end seems to have been that of 1914. The accounts of the conclave differentiated between the black smoke of the failed ballots and the white of the final one. Perhaps the principal reason for this development was Pius X's mandate that all papers relating to the election be burned, not just the ballots themselves, thus producing a great deal more white smoke for the final ballot and making it truly visible. On the first two days of the 1922 conclave, the burning of the unsuccessful ballots resulted in white smoke for some time before it turned black, creating confusion among the people. It happened again in 1939, when the white smoke changed to black only when water was thrown on the fire. Continuing problems of the same sort in 1963 resulted in the decision to use a chemical additive to make it more obvious for future conclaves.

It is difficult to prove a negative, so to maintain that no mention before 1823 of the smoke from burning the ballots demonstrates it was not used as a signal risks being disproved by a source I missed. Yet it is noteworthy that no source of the many from before 1823 that I utilized thought it of interest to mention the *sfumata*. The origins of the burning of the ballots seems lost in the fog of history, but I believe that the intentional use of the *sfumata* to reveal whether a pope has been elected is a practice at most a century old.

Notes

Chapter I

1. Some Scripture scholars raise the issue of whether the Gospels were written after the bishop of Rome had already begun to claim primacy and therefore reflect that fact rather than relating accurately Jesus's words. While a case can be made for that interpretation of the key texts, the consensus among scholars is that the Gospels predate the use of them to support claims for Roman primacy. See M. Winter, *Saint Peter and the Popes* (reprint, Westport, Conn., 1979), pp. 9–11.

2. A. Roberts and J. Donaldson, eds., *The Ante-Nicene Fathers: Translations of the Writings of the Fathers down to AD 325* (reprint, Ann Arbor, Mich., 1978), I, p. 425.

3. J. Kleist, trans., *Ancient Christian Writers* (New York, 1948), no. 6, p. 24. The work is also called the *Didache.*

4. Eusebius, *History of the Church,* trans. by G. Williamson (New York, 1966), p. 268.

5. The letters of Cyprian of Carthage, in *Ancient Christian Writers,* trans. by G. Clarke, no. 46, p. 38.

6. Two conflicting Christian sources are given in L. Loomis and J. Shotwell, *The See of Peter* (New York, 1991), pp. 629–32. The number of 137 dead comes from the pagan historian Ammianus Marcellinus, in ibid., pp. 632–33.

7. St. Jerome called Innocent the successor and son of Anastasius, but some argue he meant "spiritual son." P. Levillain, ed. *The Papacy: An Encyclopedia,* 3 vol. (New York, 2002), II, p. 781.

8. F. Noble, *The Republic of Saint Peter: The Birth of the Papal State, 680–825* (Philadelphia, 1984), pp. 188–205.

9. Before Stephen II's election a Stephen was elected who had only a four-day reign. Since he had not been ordained a bishop, the numbering system of his time did not count him as a pope. Later, the fact of election was deemed sufficient, and he was added to the list of popes; then in 1960 he was again removed

because of nonconsecration. Some lists count him as Stephen II, which complicates the numbering of the Popes Stephen who follow.

10. See Peter Stanford, *The She-Pope: A Quest for the Truth behind the Mystery of Pope Joan* (London, 1998), for a recent study of the legend that a woman who dressed as a man rose high in the Church until she was elected pope in 855. Stanford surreptitiously sat in the papal chair that supposedly was used to check if the new pope had testicles, in order to prevent a woman from ever again being elected. He found that the hole in the seat was located at precisely the right place for such a purpose. But A. Boureau, *The Myth of Pope Joan,* trans. by L. Cochrane (Chicago, 2001), argues persuasively that Pope Joan never existed.

11. O. Thatcher and E. McNeal, trans, *A Source Book for Mediaeval History* (Chicago, 1905), p. 121.

Chapter II

1. English translation in E. Henderson, trans., *Select Historical Documents of the Middle Ages* (London, 1892), pp. 361–65.

2. E. Emerton, trans., *The Correspondence of Pope Gregory VII* (New York, 1932), pp. 2–3.

3. Decree of the synod of German bishops of 1076, in Henderson, *Documents,* pp. 375–76.

4. H. Cowdrey, *The Age of Abbot Desiderius: Montecassino, the Papacy, and the Normans in the Eleventh and Early Twelfth Centuries* (Oxford, 1983), pp. 187–90.

5. See Urban's announcement of his election to the abbot of Cluny, in R. Sommerville, *Pope Urban II* (Oxford, 1996), pp. 41–42.

6. I. Robinson, *The Papacy 1073–1198* (Cambridge, 1990), pp. 68–70.

7. Quoted in ibid., p. 81. Letters by both Alexander and Victor about the disputed election of 1159 are in Thatcher and McNeal, *Source Book,* pp. 192–96.

8. Quoted in Robinson, p. 87.

9. Quoted in M. Taylor, "The Election of Innocent III," in D. Wood, ed., *The Church and Sovereignty c. 590–1918* (Oxford, 1991), p. 98.

Chapter III

1. Quoted in H. Mann, *The Lives of the Popes in the Early Middle Ages,* 20 vols. (reprint, Nendeln, Liechtenstein, 1964), XVIII, p. 29.

2. Ibid., XIX, p. 49.

3. E. Leroy Ladurie used Fournier's inquisitorial records for his best selling *Montaillou: Promised Land of Error* (New York, 1978).

4. J. Wrigley, "The Conclave and the Election of 1342," *Archivum Historicae Pontificae,* 20(1982), pp. 51–81.
5. Calculated in L. Gayer, *Le Grand Schisme d'Occident* (Florence, 1889), p. 3.
6. Text of Gregory's bull in P. Thibault, *Pope Gregory XI: The Failure of Tradition* (Lanham, Maryl., 1986), pp. 181–83.

Chapter IV

1. L. Elliott-Binns, *The History of the Decline and Fall of the Medieval Papacy* (Hamden, Conn., 1967), p. 184.
2. Quoted in L. Loomis, *The Council of Constance* (New York, 1961), p. 4.
3. Chronicle of Ulrich Richental, in ibid., p. 199. Richental was a notary from Constance.
4. Ibid., p. 140.
5. Diary of Cardinal Fillastre, in ibid., p. 428. Fillastre, a Frenchman, was one of the two cardinals who acceded to Colonna and made him pope.
6. L. von Pastor, *History of the Popes from the Close of the Middle Ages,* trans. by F. Antrobus, 40 vols. (reprint, Nendeln, Liechtenstein, 1969), I, p. 302n. This work is a major source for all the conclaves from 1417 to 1775. For some it is the only one available in English.
7. Aeneas Piccolimini (Pius II), *Memoirs of a Renaissance Pope: The Commentaries of Pius II,* trans. by F. Gragg (New York, 1959), pp. 65–66. Pius's memoirs have detailed information for this election and the next one, in which he was elected pope.
8. This quotation and those in the next paragraphs come from ibid., pp. 80–90.

Chapter V

1. Pastor, *History of Popes,* IV, p. 29.
2. Ibid., IV, 505–07, for the report of the ambassador of Milan with the list of what cardinal voted for whom. It is not arranged by ballot but by the total count of votes for the entire conclave. The best estimate is that there were three ballots.
3. D. Chambers, "Papal Conclaves and Prophetic Mystery in the Sistine Chapel," *Journal of the Wartburg and Courtauld Institutes,* 41(1978), pp. 322–26.
4. J. Burchard, *Diarium,* ed. by E. Celeni, 2 vols. (1911–42), I, pp. 30–43.
5. The next paragraphs are from F. Baumgartner, *Louis XII* (New York, 1994), pp. 179ff.
6. Sources for the conclaves of 1503 include Burchard, *Diarium,* II, pp. 245ff; M. Sanuto, *I Diarii,* ed. by F. Stefani, 58 vols. (Venice, 1879–1903),

V, pp. 73ff; F. Guicciardini, *History of Italy,* trans. by A. Goddard, 10 vols. (London, 1753–56), III, pp. 238ff; and A. Giustinian, *Dispacci,* ed. by P. Villari, 3 vols. (Florence, 1876), II, pp. 175–78. The quote is in Giustinian, II, p. 177.

7. Wagering on the papal election was an old practice by 1503, but the Venetian ambassadors' reports in Sanuto, *I Diarii,* are the first sources on the odds.

8. Giustinian, *Dispacci,* II, p. 181.

9. Description of the conclave in Archivo Segreto Vaticano, Conclavi di vari Pontifici, da Pio II a Innocente X, fol. 132 ff. This volume has capitulation texts and tally sheets for the 22 conclaves between 1503 and 1644.

10. The English ambassador stated that with three more votes Serra would have been elected, "contrary to the intentions of those who voted for him." R. Brodie, ed., *Letters and Papers, Foreign and Domestic, of the Reign of Henry VIII* (London, 1920), I, p. 770

11. D. Chambers, *Cardinal Bainbridge in the Court of Rome 1509–1514* (Oxford, 1965), p. 43.

12. Sanuto, *I Diarii,* XVI, p. 62.

13. *Letters and Papers of Henry VIII,* I, p. 798.

14. B. Hallman, *Italian Cardinals, Reform, and the Church as Property* (Berkeley, Calif., 1985), p. 66. The author notes that the tiara was mortgaged again at one of the conclaves of 1555.

15. G. Bergenroth, *Calendar of Letters . . . Relating to the Negotiations between England and Spain* (reprint, Nendeln, Liechtenstein, 1969), II, p. 383. The list of cardinals present for the conclave is given on pp. 388–90.

16. *Letters and Papers of Henry VIII,* I, p. 841.

17. *Calendar of Letters, Spain,* II, p. 586. The ambassador added that the cardinals treated Adrian not as pope but as "a private man who is on the rack."

18. *Letters and Papers of Henry VIII,* I, pp. 1416–17.

Chapter VI

1. *Calendar of Letters, Spain,* V, p. 260.

2. See J. O'Malley, *Trent and All That* (Cambridge, Mass., 2000), for a discussion of the various terms used for the study of Catholicism between 1517 and 1789. While he prefers the term "Early Modern Catholicism," his definition of Counter Reformation, on pp. 126–28, fits best for the papacy of the half-century after 1534.

3. *Calendar of State Papers and Manuscripts, relating to English affairs: Existing in the archives and collections of Venice,* 38 vols. (London, 1864), IX, p. 298.

4. For the conclave of 1549–50, see Baumgartner, "Henry II and the Papal Conclave of 1549," *The Sixteenth Century Journal,* XVI (1985), pp. 301–14; and T. Mayer, *Cardinal Pole in European Context* (Burlington, Ver., 2000).

5. *Calendar of State Papers, Venice,* IX, p. 281.

6. Quoted in Baumgartner, "Henry II and the Papal Conclave," p. 313.
7. Quoted in G. Visceglia, "Factions in the Sacred College in the Sixteenth and Seventeenth Centuries," in G. Signorotto, ed. *Court and Politics in Papal Rome, 1492–1700* (Cambridge, 2002), p. 105n. The distinction between adoration and acclamation is unclear, but it appears that adoration took place at the beginning of a conclave before any written ballot, while acclamation could occur any time after the first scrutiny.
8. *Calendar of Letters, Spain*, XIII, p. 171. See also A. Santosuosso, "An Account of the Election of Paul IV to the Pontificate," *Renaissance Quarterly*, 31(1978), pp. 486–98, for the report on the conclave by Cosimo I de'Medici's agent.
9. *Calendar of Letters, Spain*, XIII, p. 187.
10. T. Dandelet, *Spanish Rome, 1500–1700* (New Haven, Conn., 2001), p. 59. This work describes at length Spanish interference in the papal elections from 1559 to 1624.
11. See Pastor, *History of the Popes*, XV, pp. 381–90, for the balloting.
12. M-L. Rodèn, *Church Politics in Seventeenth-Century Rome: Cardinal Decio Azzolino, Queen Christina of Sweden, and the* Squadrone Volante (Stockholm, 2000), p. 168.
13. V. Pirie, *The Triple Crown* (np, 1935), pp. 105–06.

Chapter VII

1. T. Trollope, *The Papal Conclaves* (London, 1876), p. 260.
2. *Calendar of State Papers, Italy*, VIII, pp. 304–05; Pastor, *History of the Popes*, XXI, p. 239n.
3. In 1592 annual papal revenues were calculated at 1,601,200 *scudi*, so the conclave consumed about sixteen percent of the revenues, not including the pope's funeral. P. Partner, "Papal Financial Policy in the Renaissance and Counter-Reformation," *Past and Present*, 88(1980), p. 49.
4. *Traicte sommaire de l'election des papes* (Paris, 1605), folio C 1.
5. Cardinal Azzolino's *Aforismi Politici*, cited by Rodèn, *Church Politics*, p. 168.
6. On the means by which popes enriched their families, see W. Reinhard, "Papal Power and Family Strategy in the Sixteenth and Seventeenth Centuries," in A. Birke, *Princes, Patronage, and the Nobility: The Court at the Beginning of the Modern Age* (Oxford, 1991), pp. 329–58. On papal marital policy, see I. Fossi and M. Visceglia, "Marriage and politics at the papal court in the sixteenth and seventeenth centuries," in T. Dean and K Lowe, eds., *Marriage in Italy 1300–1650* (Cambridge, 1998), pp. 197–224.
7. Visceglia, "Factions in the Sacred College," p. 106.
8. G. Signorotto, "The *Squadrone Volante*: 'Independent' Cardinals and European Politics in the Second Half of the Seventeenth Century," in Signorotto, *Court and Politics*, p. 176.

9. Tally sheets for most conclaves from 1623 to 1846 are preserved in the Archivo Segreto Vaticano, Collegio di Cardinali, Conclavi. It is not clear in most cases whether they are originals or copies.

Chapter VIII

1. J. Bargrave, *Pope Alexander the Seventh and the College of Cardinals*, ed. by J. Robinson (reprint, London, 1968), article F. Barberini. Bargrave was an exiled English divine who lived in Italy for two decades. His work describes the 68 cardinals of 1670. Bargrave's work was part of a vast literature on the papacy and the conclave appearing during this era in England and throughout Europe. The fascination of Protestants and Catholics alike with the conclave reflected not only the importance of the papacy in European affairs, but also the fact that it was the only election for which the outcome was uncertain, that of the Holy Roman emperor from the Habsburg family being a foregone conclusion by this time.
2. G. Ernst, "Astrology, religion and politics in Counter-Reformation Rome," in S. Pumfrey, et al., eds., *Science, Culture and Popular Belief in Renaissance Europe* (Manchester, 1991), p. 266.
3. Bargrave, *Alexander the Seventh*, article Sacchetti.
4. Ibid., article de'Medici.
5. Signorotto, "The *Squadrone Volante*," p. 181. Initially they were called the "Swiss Cantons" because of their isolation from the factions, but the term "Flying Squadron" was adopted from a military unit that "went here and there to render aid."
6. Quoted in B. de Bildt, "The Conclave of Clement X (1670)," *Proceedings of the British Academy* (1903), p. 5.
7. Biblioteca Apostolica Vaticana, Fondo Barberini 4440. Bildt, "Conclave of Clement X," pp. 15–30, provides a far more complicated scenario for Vidoni's exclusion, which he blames largely on Azzolino's intrigue to get him elected, an example of how too much zeal in behalf of a *papabile* could undermine his chances. Bildt's article is based mostly on letters between Azzolino and Christina, which they exchanged as often as four times a day while the conclave was ongoing. Conclave officials apparently were willing to overlook their breach of the rules. See also Rodèn, *Church Politics*, pp. 217–26.
8. Pastor, *History of the Popes*, XXXI, p. 479.
9. Pirie, *The Triple Crown*, p. 193.
10. The French ambassador Brosses described the conclave in his *Journal d'Italie*, 2 vols. (Grenoble, 1971), II, pp. 234ff.
11. Pastor, *History of the Popes*, XXXVI, p. 153.
12. Ibid., XXXVI, p. 175; XXXVIII, p. 9.

Chapter IX

1. F. Gendry, "Le conclave de 1774–1775," *Revue des questions historiques,* 51(1892), pp. 424–85.
2. E. Consalvi, *Mémoires,* ed. by J. Crétineau-Joly, 2 vols. (Paris, 1866), I, pp. 230ff.
3. Ibid., p. 57
4. A. Pennington, *The Conclave* (London, 1898), p. 37.
5. G. Zizola, *Quale Papa? Analisi della Strutture Elettorali e Governative del Papato* (Rome, 1977), p. 119.
6. Comte de Chateaubriand, *Journal d'un conclave,* ed. by L. Thomas (Paris, 1913). Chateaubriand was French ambassador at Rome in 1829.
7. R. Peyrefitte, *Les Secrets des Conclaves* (Paris, 1968), p. 27.
8. N. Wiseman, *Recollections of the Last Four Popes* (London, 1858), p. 330.
9. According to several sources, Cappellari received thirty-two votes, but the conclave tally sheets in the Archivo Segreto Vaticano, Acta Conclavis Gregorii XVI, show he had thirty-three.
10. Zizola, *Quale Papa,* p. 127.
11. J. Lees-Milne, *Saint Peter's* (London, 1967), p. 318.
12. The London *Times,* July 31, 1903. The *Times* had a lengthy description of the 1878 conclave as background to the upcoming conclave.
13. The New York *Times,* August 3, 1878.

Chapter X

1. A. de Waal, *Life of Pius X* (Milwaukee, 1904), pp. 10–49. This was the first conclave for which reporters were present in large numbers. See The New York *Times,* August 1–5, 1903.
2. W. O'Connell, *Recollections of Seventy Years* (Boston, 1934), pp. 337–38.
3. G. Zizola, *Il conclave storia e segreti: l'elezione papale da San Pietro a Giovanni Paolo II* (Rome, 1993).
4. Most studies of this conclave state that the cardinal who asked for the check is unknown, but J. Pollard, in *The Unknown Pope: Benedict XV* (London, 1999), p. 62, says that it was De Lai.
5. The New York *Times,* January 31 - February 2, 1922.
6. F. Sweeney, "Cardinal O'Connell and the Conclave," *America,* 139(1978), pp. 382–83. A letter in ibid., p. 485, argues that Cardinal Dougherty of Philadelphia, the other American who missed the conclave, persuaded Pius XI.
7. F. Burkle-Young, *Papal Elections in the Age of Transition, 1878–1922* (Oxford, 2000), p. 149. Merry del Val or an ally allegedly went to Ratti before he was elected to offer his faction's support if Ratti promised not to make Gasparri the secretary of state.

8. Quoted in R. Anderson, *Between the Wars: The Story of Pius XI* (Boston, 1977), p. 53.

9. Quoted in A. Rhodes, *The Vatican in the Age of Dictators* (London, 1973), p. 219.

10. C. Falconi, *The Popes in the Twentieth Century From Pius X to John XXIII,* trans. by M. Grindrod (Boston, 1967), p. 252.

11. For a balanced assessment of Pius's critics and defenders, see J. Sanchez, *Pius XII and the Holocaust: Understanding the Controversy* (Washington, D.C., 2002).

12. F. Burkle-Young, *Passing the Keys,* 2nd edition (Lanham, Maryl., 2001), p. 67.

13. Odds given in Milan at the beginning of the conclave, cited in P. Johnson, *Pope John XXIII* (Boston, 1974), p. 110 and note. Although it is clear that betting on papal elections continued after the bull excommunicating anyone involved in such bookmaking, this is the first mention of odds since before 1800.

14. John XXIII, *Letters to His Family,* trans. by D. White (New York, 1981), p. 818.

15. Quoted in P. Hebblethwaite, *Pope John XXIII* (Garden City, N.Y., 1985), p. 296.

16. Ibid., p. 282.

17. A. Mellioni, "Pope John XXIII: Open Questions for a Biography," *The Catholic Historical Review,* 72(1986), p. 64n. See R. Flamini, *Pope, Premier, President* (New York, 1990), pp. 147–51, for a CIA memorandum to President Kennedy on the upcoming conclave.

18. Zizola, *Quale Papa,* pp. 160–72. Zizola, an Italian journalist, had many friends among the cardinals and conclavists, and he gave the vote tallies for John XXIII with some certainty. He is less sure of the numbers for this one.

19. Flamini, *Pope,* pp. 173–74.

20. The English text of Paul's decree is in Burkle-Young, *Passing the Keys,* pp. 467ff.

21. The account of the voting by F. X. Murphy in *Newsweek,* September 11, 1978, is regarded as the most authoritative. Murphy had become friends with many cardinals while writing about Vatican II. G. Thomas and M. Morgan-Witts, *Pontiff* (Garden City, N.Y.), 1983, p. 193, maintain that Luciani had only sixty-six votes in the third scrutiny.

22. See D. Yallop, *In God's Name: An Investigation into the Murder of Pope John Paul I* (Toronto, 1984), for the case that he was murdered. See J. Cornwell, *A Thief in the Night: The Death of Pope John Paul I* (London, 1990), for a rebuttal.

Chapter XI

1. Quoted in M. Hebblethwaite, *The Next Pope* (New York, 2000), p. 70.

2. A complete list of the cardinals with their birth dates, dates of appointments, and homelands is online, Salvador Miranda, http://www.fiu.edu/~mirandas/cardinals.htm.

3. Its text can be found in English in Burkle-Young, *Passing the Keys,* pp. 457–65, and online at http/.www.newadvent.org/docs/jp02ud.htm.

4. There are currently four men who proclaim themselves to be pope, who have at least a few followers. Gregory XVII, a Spaniard, claims the papacy on the basis of a revelation from Christ. He accepts Paul VI as valid but refers to the two John Pauls as villains. Clement XV, a Canadian, received a message from heaven in 1950 that he would become pope and claims now to be the successor to John XXIII. He is at least ninety years old. Michael I, based in Kansas, denounces John XXIII and his successors as apostates. He claims that the papacy was vacant from 1958 to 1990, when he was elected by the faithful remnant of the Church. Pius XIII, another American, likewise proclaims himself to be the legitimate successor to Pius XII, whom he considers the last true pope. He claims that he was elected in a conclave held in Montana on October 23, 1998. Lacking the funds to bring the electors to Montana, the election was held by telephone. Otherwise that conclave was conducted as closely as possible to the standard procedure, including the white smoke from burning the ballots.

5. M. Hebblethwaite, *The Next Pope,* pp. 120–27.

Appendix

1. Pastor, *History of the Popes,* XXX, p. 23.

2. "The Diary of Cardinal Fillastre," in Loomis, *The Council of Constance,* p. 427.

3. *Traicte sommaire de l'election des papes* (Paris, 1605), folio C 1. On the use of the pyramid-shaped stove, see Rodèn, *Church Politics,* p. 167. On the addition of the chimney, see L. Lector, *L'élection papale* (Paris, 1896), p. 263. He states: "This ritual method of burning the ballots dates to a little before the middle of the last century. Originally they were burned in a simple ceremony in the same hall as the voting. . . . This unique ceremony arose from the idea of installing the small chimney visible from the outside." Since he was writing in the nineteenth century, presumably he meant the innovation occurred in the eighteenth century. Thomas, *Pontiff,* p. 182, proposes that the chimney was introduced by Julius III (1550–1555) because of damage to the frescoes, but Pastor makes no mention in his volume on Julius (vol. XIII) of such a decision, despite a long discussion of Julius's relationship with Michelangelo. Furthermore, the balloting in the conclave of 1549–50 took place in the Pauline Chapel. See above, p. 105.

4. G. Leti, *The Ceremonies of the Vacant See, or, A true relation of what passes at Rome upon the pope's death with the proceedings in the conclave, for the election*

of a new pope, trans. by J. Davies (London, 1671), p. 72. Leti's original Italian edition was written shortly after the conclave of 1667.

5. Ibid., p. 86. *A New History of the Roman Conclave* (London, 1691), p. 17, says much the same.

6. F. Gendry, "Le conclave de 1774–1775," *Revue des questions historiques,* 51(1892), pp. 424–85.

7. R. De Cesare, *Il Conclave di Leone XIII con aggiunte e nouvi documenti* (Citta di Castello, 1888), p. 330. See also the London *Times,* September 2, 1823, p. 2.

8. B. O'Reilly, *Pius IX* (np, 1877), p. 79.

9. The New York *Times,* February 20, 1878. The story on the conclave from February 19 states: "The smoke of burning ballots was visible at 1:15 this afternoon showing that the Conclave had voted but nobody had obtained the necessary majority." This account of the use of the smoke only to announce a failed ballot is found also in Trollope, *Papal Conclave,* p. 413; and Pennington, *The Conclave,* p. 71.

10. The London *Times,* July 31, 1903; J. Keller, *The Life and Acts of Leo XIII* (St. Louis, 1879), p. 246. Keller notes the burning of the ballots at several points but makes no mention of the color of the smoke.

11 The New York *Times,* August 2, 1903

Bibliography

(Items cited in the notes are not repeated here.)

General Works

Z. Aradi. *The Popes: The History of How They are Chosen, Elected, and Crowned.* London, 1956.

P. Bander. *The Prophecies of St. Malachy.* Exeter, England, 1979.

J. Broderick. "The Sacred College of Cardinals: Size and Geographical Composition (1099–1986)." *Archivum Historicae Pontificiae,* 25(1987), pp. 7–71.

N. Cheetham. *Keepers of the Keys: A History of the Popes from St. Peter to John Paul II.* New York, 1983.

E. Duffy. *Saints and Sinners: A History of the Popes.* New Haven, 1997.

C. Eubel, et al. *Hierarchia Catholica medii et recentioris aevi.* 8 vols. Münster, Germany, 1918–1936.

F. Gontard. *The Chair of Peter: A History of the Papacy.* Trans. by A. and E. Peeler. New York, 1964.

H. Jedin, et al. *The History of the Church.* 10 vols. London, 1965–81.

P. Johnson. *The Papacy.* London, 1997.

J. Kelly. *The Oxford Dictionary of the Popes.* Oxford, 1986.

K. Schatz. *Papal Primacy: From its Origins to the Present.* Trans. by J. Otto. Collegeville, Minn., 1996.

B. Schimmelpfenning. *The Papacy.* New York, 1992.

G. Zizola. *Il Conclave Storia e Segreti: L'elezione papale da San Pietro a Giovanni Pietro II.* Rome, 1993.

Chapter I

G. Alberigo and A. Weiler, eds. *Election and Consensus in the Church.* New York, 1972.

J. Alchermes, "Petrine Politics: Pope Symmachus and the Rotunda of St. Andrew at Old St. Peter's," *Catholic Historical Review,* 81(1995), pp. 1–40.

O. Culmann. *Peter: Disciple, Apostle, Martyr.* London, 1962.

R. Davis, trans. *The Book of Pontiffs (Liber Pontificalis): The Ancient Biographies of the First Ninety Popes to AD 715.* Liverpool, 2000.

R. Davis, trans. *The Lives of the Eighth-century Popes: The Ancient Biographies of Nine Popes from AD 715 to AD 817.* Liverpool, 1992.

R. Eno. *The Rise of the Papacy.* Wilmington, Del., 1990.

E. Ferguson. "Origen and the Election of Bishops." *Church History,* 43(1974), pp. 26–33.

J. Richards. *Consul of God: The Life and Times of Gregory the Great.* London, 1980.

J. Richards. *The Popes and the Papacy in the Early Middle Ages, 476–752.* London, 1979.

D. Wood. "The Pope's Right to Elect His Successor: The Criterion of Sovereignty." In D. Wood, ed. *The Church and Sovereignty c. 590–1918.* Oxford, 1991.

Chapter II

M. Baldwin. *Alexander III and the Twelfth Century.* New York, 1968.

G. Barraclough. *The Medieval Papacy.* London, 1968.

F. Logan. *A History of the Church in the Middle Ages.* New York, 2002.

C. Morris. *The Papal Monarchy: The Western Church from 1050 to 1250.* Oxford, 1989.

P. Partner. *The Lands of St Peter: The Papal State in the Middle Ages and the Early Renaissance.* London, 1972.

K. Pennington. *Pope and Bishops: The Papal Monarchy in the Twelfth and Thirteenth Centuries.* Philadelphia, 1984.

M. Stoll, *The Jewish Pope: Ideology and Politics in the Papal Schism of 1130.* Leiden, Netherlands, 1987.

R. Southern. *Western Society and the Church in the Middle Ages.* Grand Rapids, Mich., 1970.

W. Ullman. *A Short History of the Papacy in the Middle Ages.* London, 1972.

K. Woody. "*Sagena Piscatoris:* Peter Damiani and the Papal Election Decree of 1059." *Viator,* 1(1970), 33–54.

Chapter III

M. Creighton. *A History of the Papacy from the Great Schism to the Sack of Rome.* 5 vols. London, 1907–1911.

J. Eastman. *Papal Abdication in Later Medieval Thought.* Lewiston, N.Y., 1990.

P. Herde. "Election and Abdication of the Pope: Practice and Doctrine in the Thirteenth Century." In S. Kuttner, ed. *Proceedings of the Sixth International Congress of Mediaeval Canon Law.* Vatican City, 1985, pp. 411–36.

G. Mollat. *The Popes at Avignon.* Edinburgh, 1963.

E. Petrucelli della Gattina, *Histoire diplomatique des conclaves,* 4 vols. Paris, 1864.

L. Salembier. *The Great Schism of the West.* Reprint, London, 1968.

W. Ullman. *The Growth of Papal Government.* London, 1970.

W. Ullman. *The Origins of the Great Schism.* London, 1948.

Chapter IV

C. Black. *Council and Commune: The Conciliar Movement and the Council of Basel.* London, 1979.

J. Gill. *Eugenius IV: Pope of Union.* New York, 1982.

A. Glasfurd. *The Antipope: Peter de Luna, 1342–1423; A Study in Obstinacy.* London, 1965.

D. Hay. *The Church in Italy in the Fifteenth Century.* Cambridge, 1977.

F. Oakley. *The Western Church in the Later Middle Ages.* Ithaca, N.Y., 1979.

J. Smith. *The Great Schism.* London, 1978.

B. Tierney. *Foundations of Conciliar Theory.* Cambridge, 1955.

Chapter V

J. D'Amico. *Renaissance Humanism in Papal Rome.* Baltimore, 1983.

E. Lee. *Sixtus IV and Men of Letters.* Rome, 1978.

K. Lowe. *Church and Politics in Renaissance Italy: The Life and Career of Cardinal Francesco Soderini (1453–1524).* New York, 1993.

M. Mallet. *The Borgias.* London, 1969.

J. Sägmüller. *Die Papstwahlen und die Statten von 1447 bis 1555.* Reprint, Tübingen, Germany, 1967.

C. Shaw. *Julius II, the Warrior Pope.* Oxford, 1993.

J. Thompson. *Popes and Princes, 1417–1517.* London, 1980.

Chapter VI

A. Antonovics. "Counter-Reformation Cardinals: 1534–90." *European Studies Review.* 4(1972), pp. 301–28.

G. Kittler. *The Papal Princes.* New York, 1960.

S. Ostrow. *Art and Spirituality in Counter-Reformation Rome.* Cambridge, 1996.

P. Prodi. *The Papal Prince One Body and Two Souls: The Papal Monarchy in early Modern Europe.* Trans. by S. Haskins. Cambridge, 1987.

A. Santosuosso. "An Account of the Election of Paul IV to the Pontificate." *Renaissance Quarterly.* 31(1978), pp. 486–98.

A Wright. *The Early Modern Papacy: From the Council of Trent to the French Revolution.* London, 2000.

Chapter VII

K. Comerford and H. Pabel, eds. *Early Modern Catholicism: Essays in Honour of John W. O'Malley, S.J.* Toronto, 2001.

M. Liebman. "Journal secret d'un conclave." *La Revue Nouvelle,* 38(1963), pp. 32–52.

C. Paillat and P. Lesourd. *Dossier secret des conclaves.* Paris, 1969.

G. Signorotto, ed. *Court and Politics in Papal Rome.* Cambridge, 2002.

W. Weech. *Urban VIII.* London, 1905.

C. Wood. *Gregory XV (1621–1623).* Ann Arbor, Mich., 1990.

Chapter VIII

M. Andrieux. *Daily Life in Papal Rome in the Eighteenth Century.* Trans. by M. Fitton. New York, 1968.

W. Callahan and D. Higgs, eds. *Church and Society in Catholic Europe of the Eighteenth Century.* Cambridge, 1979.

O. Chadwick. *The Popes and European Revolution.* Oxford, 1981.

H. Gross. *Rome in the Age of Enlightenment.* Cambridge, 1990.

R. Haynes. *Philosopher King: The Humanist Pope Benedict XIV.* New York, 1970,

Chapter IX

R. Aubert. *The Church in the Industrial Age.* Trans. by M. Resch. New York, 1981.

D. Beales, ed. *History, Society, and the Churches.* Cambridge, 1985.

F. Coppa. *Pope Pius IX: Crusader in a Secular Age.* Boston, 1979.

E. Hales. *The Church and the Modern World.* London, 1958.

E. Hales. *The Emperor and the Pope: The Story of Napoleon and Pius VII.* New York. 1978.

J. Schmidlin. *Histoire des papes de l'époque contemporaine,* 2 vols., Paris, 1940.

C. Terlinden. "Le conclave de Léon XII (2–28 septembre, 1823)." *Revue d'histoire ecclésiastique.* XIV(1913).

Chapter X

A. Greeley. *The Making of the Popes 1978: The Politics of Intrigue in the Vatican.* Kansas City, Kansas, 1979.

H. Grissell de La Garde. *Sede Vacante, being a Diary Written during the Conclave of 1903.* Oxford, 1903.

P. Hebblethwaite. *Paul VI: The First Modern Pope.* New York, 1993.

P. Hebblethwaite. *The Year of the Three Popes.* London, 1978.

Chapter XI

J. Allen. *Conclave.* New York, 2002.

T. Reese. *Inside the Vatican: The Politics and Organization of the Catholic Church.* Cambridge, Mass., 1996.

T. Szulc. *Pope John Paul II.* New York, 1995.

M. Walsh. *John Paul II.* London, 1994.

Index